FIBRE OPTICS
Theory and Practice

OPTICAL PHYSICS AND ENGINEERING

Series Editor: William L. Wolfe
Optical Sciences Center
University of Arizona
Tucson, Arizona

1968:
M. A. Bramson
Infrared Radiation: A Handbook for Applications

1969:
Sol Nudelman and S. S. Mitra, Editors
Optical Properties of Solids

1970:
S. S. Mitra and Sol Nudelman, Editors
Far-Infrared Properties of Solids

1971:
Lucien M. Biberman and *Sol Nudelman, Editors*
Photoelectronic Imaging Devices
   Volume 1: Physical Processes and Methods of Analysis
   Volume 2: Devices and Their Evaluation

1972:
A. M. Ratner
Spectral, Spatial, and Temporal Properties of Lasers

1973:
Lucien M. Biberman, Editor
Perception of Displayed Information
W. B. Allan
Fibre Optics: Theory and Practice ·

FIBRE OPTICS
Theory and Practice

by

W. B. Allan
Ministry of Defence
Fort Halstead
Sevenoaks
Kent

Springer Science+Business Media, LLC 1973

Library of Congress Catalog Card Number: 72–95066

ISBN 978-1-4684-2042-5   ISBN 978-1-4684-2040-1 (eBook)
DOI 10.1007/978-1-4684-2040-1

Preface

The emergence of fibre optics as a commercially viable technology occurred barely ten years ago; in this time it has become an established field with a variety of applications. This book has been written in an attempt to review the entire field with an emphasis on the practical applications of the technology. This approach has been adopted since it was felt that there was a need for a work which could be referred to by non-specialists in the field who were interested in, or who wished to make use of, fibre optics.

With this readership in mind, the theory has been presented in as simple a manner as possible and emphasis has been placed on the description of typical applications and the manufacturing techniques of the technology. It is hoped that this mode of presentation will enable the reader to form an appreciation of both its advantages and its limitations.

In Chapter 1, a brief historical introduction is presented of the technology. In Chapters 2 and 3, the theory and practice of the optical fibre is outlined. Chapters 4 and 5 describe the manufacture and applications of the simplest type of component met in fibre optics, namely the light guide. Chapters 6, 7 and 8 provide, respectively, the theory, manufacture and applications of coherent bundles of optical fibres. Chapter 9 deals with the optical fibre regarded as a waveguide for electromagnetic energy. Chapter 10 is a pot-pourri of topics which are of secondary importance (at present!), and includes: fibre optics outside the visible spectrum, active optical fibres and optical fibres with a graded index. Chapter 11 is a review of the technology as it is applied in the medical field.

It is always difficult in a work of this nature to single out people who have influenced the author, since many of the developments described were team efforts; however, I would like to acknowledge the help and advice which I received from Mr J. M. Ballantine and Dr A. J. Worral, both of Barr and Stroud Ltd, during my stay with that firm. In addition, I had many stimulating discussions with the former regarding the structure of a book on fibre optics, and he has also kindly read and commented on the manuscript. I also feel that I must acknowledge the encouragement I received from Dr T. Flitcroft while he was General Manager at Rank Kershaw (Rank Precision Industries Ltd), during which time a number of commercially successful developments were initiated, some of which are described in the book. Finally, I must thank my wife who undertook the arduous task of typing and correcting the manuscript, and without whose moral support the book would not have been written.

W. B. Allan

Contents

Contents

CHAPTER 1

Historical Introduction

The field of fibre optics concerns itself with the guidance of light by multiple reflections along channels formed from glass or plastic. The reflection process which is invariably employed is that of total internal reflection at a dielectric interface, and the first recorded observation of this principle was by John Tyndall[1] in 1854, who, in a lecture to the Royal Institution in London, "... permitted water to spout from a tube, the light on reaching the limiting surface of air and water was totally reflected and seemed to be washed downwards by the descending liquid ...". In this case, the dielectric interface was formed between water and air, the air being the material of lower refractive index.

It was not until the early twentieth century that a practical application of this phenomenon was proposed. In 1927, Baird[2] in the UK, and Hansell[3] in the USA applied for patents for image-transferring devices using fibres of silica. In the 1930s illumination was provided, in medical inspection instruments in the UK by using a plastic channel to guide the light to the inspection area; and in Germany, Lamm[4] published a description of the use of flexible fibres for the transmission of images in gastroscopy. However, since uncoated fibres were used, the efficiency was low; therefore these ideas were not actively pursued and lay dormant until the early 1950s. At that time, the simultaneous publication of papers by van Heel[5] in Holland and Hopkins and Kapany[6] in the UK, both dealing with image transmission along bundles of fine glass fibres, caused a resurgence of interest in this field which has been maintained to the present day. It was van Heel who pointed the way towards the high reflection efficiencies required in fibre optics, by coating his fibres with a solid sheath of lower refractive index, thereby protecting the reflecting surface. The coating used by van Heel was a plastic material and not altogether satisfactory, but the development of glass coatings was reported in 1958 by Hirschowitz *et al.*[7] Courtney-Pratt in 1954 proposed the use of optical fibres, in the form of fused plates, for the screens of electron tube devices.

The major part of the development of fibre optics occurred, during this period, in the USA through the work of Kapany and that of the American Optical Company. Outside America, work was started in 1957 by Coleman at the Atomic Weapons Research Establishment,

Aldermaston and in 1959 by the author at Barr & Stroud Ltd, both in the UK. By the early 1960s most of the basic development had been completed, and the major emphasis since then has been on applications of the technology.

The milestones in the development of this technology have been:

(a) The introduction of the solid sheath by van Heel.

(b) The development of "multiple-fibres" by Kapany[8] in 1958.

Process (b) involved the production of very fine fibres fused together as a mechanical unit; the techniques used are similar to those employed in Palestine to produce glass mosaics in the first century B.C.![9]

(c) The introduction by the American Optical Company of a second, absorbing sheath to reduce the effect of stray light.

(d) The development by Snitzer[10] of the optical-fibre laser, in 1961.

(e) The development by the Nippon Glass Company of Japan[11] of a graded index fibre, in 1968.

The technology of fibre optics is utilized in two broad areas of application: the transport of light and the transport of visual information. The first area uses bundles of optical fibres in which the fibres are not aligned and these are known as non-coherent bundles, or more simply, light guides. In the second area the fibres are aligned and the resulting bundles are called coherent. This latter area can be conveniently divided into two groups, one of which utilises a flexible bundle, and the other a solid (rigid) bundle, since the manufacturing techniques adopted are completely different. In the non-coherent area, the main commercial applications have been the provision of illumination for medical instruments and the sensing of holes in punched-card readers. In the coherent area, the flexible bundle has found its main application in medical inspection, whilst the solid bundle has been used almost exclusively as a faceplate in electron optical image tubes.

It will be seen that, although the basic principle underlying the whole of fibre optics was demonstrated about 120 years ago, the main development activity started less than 20 years ago and the commercial exploitation of this technology is barely 10 years old. In this comparatively short time, the technology has been established as a viable field of optics, with a range of applications which are undeniably valid; in fact some of them cannot be fulfilled in any other way. This is perhaps the test of any new technology, that, when the novelty has worn off and the far-fetched applications have been rooted out, there remains a hard core of valid applications which justify its existence. This has certainly happened with fibre optics, and the indications are that the list of valid applications will increase over the next few years.

REFERENCES

1. J. Tyndall, *Proc. Roy. Instn.* **1**, 446 (1854)
2. J. L. Baird, *Brit. Appl.* 20,969/27, *Brit. Patent* 285,738 (1927)
3. C. W. Hansell, *US Patent* 1,751,584 (1930)
4. H. Lamm, *Z. Instrumentenk.* **50**, 579 (1930)
5. A. C. S. van Heel, *Nature* **173**, 39 (1954)
6. H. H. Hopkins and N. S. Kapany, *Nature* **173**, 39 (1954)
7. B. I. Hirschowitz, L. E. Curtiss, C. W. Peters and H. M. Pollard, *Gastroenterology* **35**, 50 (1958)
8. N. S. Kapany, *J. Opt. Soc. Am* **49**, 779 (1959)
9. A. C. Revi, *Glass Industry* **38**, 328 (1957)
10. E. Snitzer, *J. Appl. Phys.* **32**, 36 (1961)
11. T. Uchida, M. Furukawa, I. Kitano, K. Kojumi and H. Matsumara, Paper presented at the IEEE/Opt. Soc. Am. Joint Conf. Laser Eng. Appl., May 1969, Washington, D.C.

CHAPTER 2

Fibre Optics – Basic Theory

The fundamental unit in any fibre optics system is the individual optical fibre, whose properties will determine the performance of the whole system. In this chapter the theoretical performance of the fibre will be discussed, and used in later chapters to describe and assess the performance of combinations of optical fibres.

An optical fibre is basically a guidance system which is cylindrical in shape. If a beam of electromagnetic energy enters this system through one end-face of the cylinder, then a significant portion of this energy, usually light, will be trapped within the system and guided through it to emerge from the other end-face. Guidance is achieved by causing the beam to be multiply reflected at the cylinder walls, and, since the beam undergoes a large number of reflections, total internal reflection is used to provide a high reflectivity. Therefore an optical fibre can be defined as a cylinder of transparent dielectric material, of refractive index n_1, whose walls are in contact with a second dielectric material of a lower refractive index n_2. For the sake of generality it will be assumed that the fibre is immersed in a third dielectric material of refractive index n_0, although this medium is normally air ($n_0 = 1.0$).

The theory will be developed in two stages. In the first of these will be considered the behaviour of a ray of light whose passage through the fibre lies in a single plane. This is defined as a meridional ray and is the simplest to describe mathematically. In the second stage, the behaviour of a ray whose path is not confined to a single plane will be discussed, such a ray being defined as a skew ray. In order to avoid complicating the theory unduly, it will be assumed in the above that the absorption, reflection and Fresnel losses are zero. The reduction in efficiency caused by these losses when finite will be treated in a later chapter. Further, the size of the fibre will be assumed to be large compared to the wavelength of light used, and the treatment will be based on geometrical optics. Finally, we shall consider the phenomenon of total internal reflection from a physical standpoint, and the effects of this on the propagation of light through the fibre will be examined.

I. TOTAL INTERNAL REFLECTION

Guidance is achieved in an optical fibre by causing the light to be internally reflected at the cylinder walls, this type of reflection being chosen for its high efficiency. A ray of light will be internally reflected at the boundary between two dielectric media when the ray is incident within the denser medium, and the angle of incidence is greater than a critical value defined by the refractive indices of the media. The critical angle for internal reflection can be calculated from Snell's law. In Fig. 1 a light beam is incident at the boundary between two media defined by refractive indices n_1 and n_2 ($n_1 > n_2$). The ray reaches the interface through the denser medium, and the angle of refraction θ_2 is related to the angle of incidence θ_1 by Snell's law, which states that:

$$n_2 \sin \theta_2 = n_1 \sin \theta_1$$

or

$$\sin \theta_2 = \frac{n_1}{n_2} \sin \theta_1 \tag{1}$$

Since $n_1 > n_2$, $\sin \theta_2 > \sin \theta_1$, so that $\sin \theta_2$ equals unity for a value of θ_1 which is less than 90°. For values of θ_1 greater than this, Eq. (1) will yield a value for $\sin \theta_2$ greater than unity, which is mathematically impossible. Physically, this paradox is resolved by the ray being internally reflected for such values of θ_1. The critical angle for reflection (θ_{ic}) is therefore derived from Eq. (1) by setting $\sin \theta_2$ equal to unity, which gives:

$$\sin \theta_{ic} = n_2/n_1 \tag{2}$$

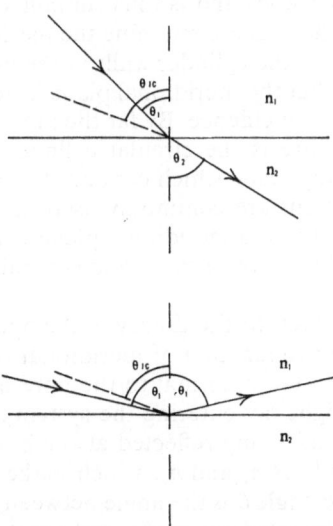

Fig. 1   Refraction and reflection of light rays at a dielectric interface defined by refractive indices n_1 and n_2 ($n_1 > n_2$).

A ray of light will be reflected at the boundary for all angles of incidence greater than θ_{ic}.

Clearly the above argument only applies when $n_1 > n_2$, i.e. the ray meets the boundary from the denser medium. In order to explain this phenomenon, electromagnetic theory must be applied to the above situation. However, this will be postponed to a later section, since the theory of the optical fibre can be developed with the knowledge of the reflection condition alone.

The phenomenon of internal reflection has been known for a long time, and its use in folded and inverting optical systems is commonplace. It provides an extremely efficient reflection system. Potter[1] has estimated, from measurement of optical fibers, values of reflectivity for glass – glass boundaries of 0.9995. The losses are thus several orders of magnitude lower than for metallic reflection; aluminium exhibits a reflectivity of around 0.9. The system is therefore an ideal choice for any application which involves a large number of reflections.

II. MERIDIONAL RAYS

A. Propagation of Meridional Rays

A meridional ray is one whose path through the fibre is confined to a single plane, and the mathematical treatment of the propagation is reduced to two dimensions and is thus simplified. It is obvious from this definition that the plane containing the meridional ray must also contain the normals to the cylinder walls at the points of incidence of the ray. This means that the meridional plane is normal to the tangent planes at the points of incidence. By far the most common configuration for an optical fibre is the circular cylinder, and, in this case, a meridional plane is any plane which contains the cylindrical axis. Only two other configurations are commonly used: the hexagonal and the square cylinder. For these a meridional plane is any plane normal to a pair of diametrically opposite sides, and is parallel to the cylindrical axis.

The simplest approach to the theory of the optical fibre entails the consideration of the propagation of meridional rays along a straight optical fibre with end-faces normal to the fibre axis. Figure 2 shows such a fibre with a light ray entering the system from a medium with refractive index n_0 and being reflected at the boundary between the media, defined by indices n_1 and n_2, which make up the optical fibre system. The entrance angle θ is the angle between the ray and the axis in medium n_0, θ' is the refracted angle, and ϕ is the angle of incidence at the dielectric boundary.

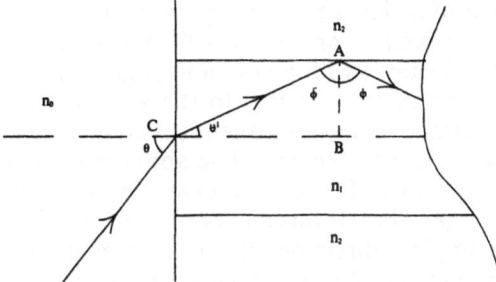

Fig. 2   The passage of a light ray into an optical fibre.

From Snell's law:

$$n_0 \sin \theta = n_1 \sin \theta^1$$
$$= n_1 \cos \phi \qquad (3)$$

For reflection to occur we must have:

$$\sin \phi > n_2/n_1$$

i.e.
$$\cos \phi < (1 - n_2^2/n_1^2)^{\frac{1}{2}} \qquad (4)$$

If a substitution for inequality (4) is made in Eq. (3) the following expression is obtained:

$$n_0 \sin \theta < (n_1^2 - n_2^2)^{\frac{1}{2}} \qquad (5)$$

The inequality (5) gives the angular condition which must be satisfied by a meridional ray before internal reflection can occur. Dividing by n_0 gives the condition:

$$\sin \theta < \frac{1}{n_0} (n_1^2 - n_2^2)^{\frac{1}{2}} \qquad (6)$$

It should be noted that the right-hand side of inequality (6) is greater than unity if:

$$n_1^2 > n_2^2 + n_0^2$$

The physical interpretation of this is that light which is incident on the end-face at the maximum possible angle (i.e. normal to the axis) will be incident at the dielectric interface at an angle greater than the critical angle for total internal reflection. In this case $\sin \theta$ has a maximum value of unity.

The intersections of the dielectric boundary with the meridional plane will take the form of two parallel lines for a straight cylinder.

Thus, the angles of incidence at subsequent reflections will be equal to that at the first reflection, and a ray will be guided through the fibre by a series of reflections with a common angle of reflection, ϕ. If the output face of the fibre is normal to the axis and is immersed in a medium with refractive index n_0, then it is obvious that the emerging ray lies at an angle θ to the axis. The absolute direction of this ray depends on the number of reflections experienced by the ray during its passage through the fibre. If this number is even, then the ray emerges parallel to its original direction. If this number is odd, then the ray emerges at an angle of 2θ to its original direction. Figure 3 shows the passage of two parallel rays through a fibre with the number of reflections differing by one.

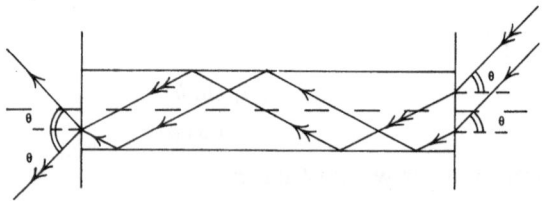

Fig. 3   The passage of two parallel light rays through an optical fibre, where the number of reflections differs by one.

Therefore a straight optical fibre, comprising media with refractive indices n_1 and n_2 $(n_1 > n_2)$, which is immersed in a medium with refractive index n_0, will accept and propagate a cone of light incident on its end-face, provided the conical semi-angle is less than θ_M, where θ_M is defined by the upper limit of inequality (6). That is:

$$\sin \theta_M = \frac{1}{n_0} (n_1^2 - n_2^2)^{\frac{1}{2}} \text{ when } n_1^2 < n_2^2 + n_0^2$$
$$= 1.0 \text{ when } n_1^2 > n_2^2 + n_0^2 \tag{7}$$

Normally, the fibre is immersed in air or in a vacuum, in which case with a sufficient degree of accuracy $n_0 = 1.0$.
Then Eq. (7) can be written:

$$\sin \theta_M = (n_1^2 - n_2^2)^{\frac{1}{2}} \text{ when } n_1^2 < n_2^2 + 1$$
$$= 1.0 \text{ when } n_1^2 > n_2^2 + 1 \tag{8}$$

Equation (8) is clearly a measure of the light-gathering power of the fibre, and the quantity given by the upper term on the right-hand side of Eq. (8) is defined as the numerical aperture (N.A.) of the fibre, by analogy with lens optics. Thus, the N.A. of any optical fibre is defined as:

$$\text{N.A.} = (n_1^2 - n_2^2)^{\frac{1}{2}} \tag{9}$$

It will be seen later that, although the above argument does not apply exactly to skew rays, Eq. (8) is a sufficiently accurate description of the collecting efficiency of an optical fibre to justify its general usage. Figure 4 is a nomogram which inter-relates n_1, n_2 and N.A. Equation (8) defines the optical performance of a straight optical fibre and exhibits a novel characteristic which distinguishes an optical fibre from a conventional optical system. This is that the maximum acceptance angle of the fibre system is defined solely by the refractive indices of the system, and is independent of its physical dimensions. There are two major consequences of this fact; firstly, the acceptance angles can be made large, i.e. the N.A. can approach, or exceed, unity and secondly, the fibre cross-section can be made small so that the fibre can become flexible. The exploitation of these features has created the field of fibre optics as it is known at present.

Path Length and Number of Reflections

Referring again to Fig. 2, it can be seen that the path length between reflections is given by $2 \times AC$, whilst the axial length between reflections is $2 \times BC$. Thus, in an optical fibre, the path length per unit length

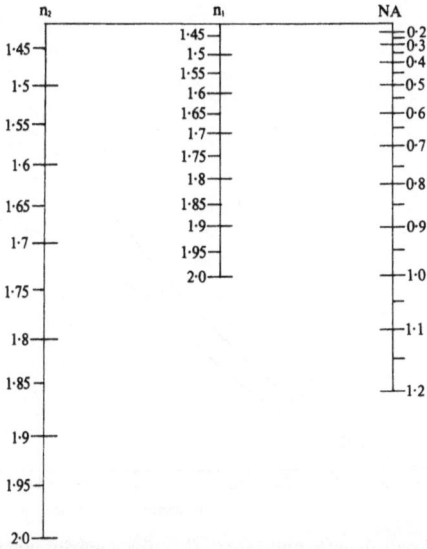

Fig. 4   Nomograph which inter-relates core index (n_1), sheath index (n_2) and numerical aperture.

(l_m) is given by:

$$l_m = AC/BC = \sec \theta^1$$

$$= \frac{n_1}{(n_1^2 - \sin^2 \theta)^{\frac{1}{2}}} \text{ when } n_0 = 1.0 \qquad (10)$$

This is independent of fibre dimensions and is controlled by the refracted angle. Figure 5 shows the variation of l_m with θ for two typical values of n_1, viz. 1.6 and 1.8.

Further, the number of reflections per unit length (η_m) is given by $1/2BC$, so that:

$$\eta_m = \frac{1}{2BC} = \frac{\tan \theta^1}{d}, \text{ where } d \text{ is the fibre diameter}$$

$$= \frac{1}{d \tan \phi} \qquad (11)$$

$$= \frac{\sin \theta}{d(n_1^2 - \sin^2 \theta)^{\frac{1}{2}}} \text{ when } n_0 = 1.0 \qquad (12)$$

Thus, η_m is inversely proportional to the fibre diameter, and Fig. 6 shows the variation of $\eta_m d$ with θ for two typical values of n_1, viz. 1.6 and 1.8.

It will be noted from Fig. 6 that a high reflectivity is essential when fibres of small diameters are being used.

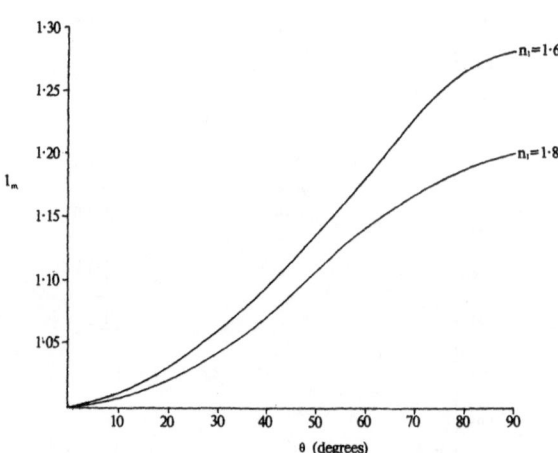

Fig. 5 Variation of path length/unit length (l_m), for a meridional ray, plotted against input axial angle (θ).

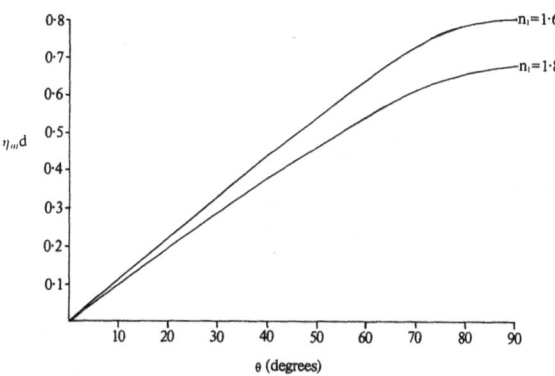

Fig. 6 Variation of number of reflection/unit length times fibre diameter (η_m d), for a meridional ray, plotted against input axial angle (θ).

B. The Light-Gathering Power of an Optical Fibre

In many applications,[2] optical fibres are used in systems which involve the collection of light from a Lambertian source, i.e. a source whose intensity is given by $I_0 \cos \theta$ where θ is the angle between the normal to the source and the direction of measurement. In these cases it is important to know the amount of light collected from such a source by an optical fibre with an N.A. equal to $\sin \theta_M$, where $\sin \theta_M$ is less than unity. A fibre with an N.A. of unity or greater will clearly collect all the light from a Lambertian source, ignoring Fresnel reflections at the end-faces and absorption losses.

The amount of light collected by a fibre which is normal to a Lambertian source of infinite extent equals that transmitted by the same fibre in contact with the source. In the latter case, it is obvious that light will only be collected from the source within an area defined by the dielectric boundary of the fibre. From any point within this area light will be collected by meridional rays up to an angle defined by equation (8). Figure 7 shows a hemisphere, radius r and centre C on a Lambertian source. Light emitted from the point C and within the angles θ and $\theta + \mathrm{d}\theta$ to the normal will pass through the shaded band. The amount of light ($\mathrm{d}F$) passing through this band is given by:

$$\mathrm{d}F = \frac{I_0 \cos \theta}{r^2} \mathrm{d}A \tag{13}$$

whose $\mathrm{d}A$ is the area of the band. Now:

$$\mathrm{d}A = r \, \mathrm{d}\theta \times 2\pi r \sin \theta$$

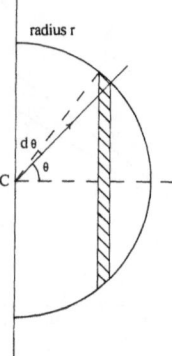

Fig. 7   Light output from a Lambertian source through a spherical shell, radius r.

which on substitution in Eq. (13) yields:

$$dF = 2\pi I_0 \sin\theta \cos\theta \, d\theta \qquad (14)$$

The amount of light collected by the optical fibre will be given by integrating Eq. (14) between $\theta = 0$ and $\theta = \theta_M$. That is:

$$F = \int_0^{\theta_M} 2\pi I_0 \sin\theta \cos\theta \, d\theta$$

$$= 2\pi I_0 \int_0^{\theta_M} \sin\theta \, (d\sin\theta)$$

$$= \pi I_0 \sin^2\theta_M \qquad (15)$$

The total flux emitted by this point is πI_0 and the fraction of light collected by the fibre, f_m, is given by:

$$f_m = \sin^2\theta_M = (\text{N.A.})^2 \text{ when N.A.} \ll 1.0$$

$$= 1.0 \text{ when N.A.} \gg 1.0 \qquad (16)$$

The above calculations show the importance of the N.A. of the fibre in relation to its light-gathering properties. Figure 8 shows the variation of f_m with N.A.; for comparison, the theoretical efficiencies of lenses of differing f numbers, working at $1:1$ conjugates, are also indicated on this graph – these have been calculated ignoring Fresnel reflections.

It will be seen from Fig. 8 that, since N.A.s of 1.0 can readily be achieved in optical fibres, lenses come a very poor second to high-N.A. fibres when considering light collection from a Lambertian source.[3]

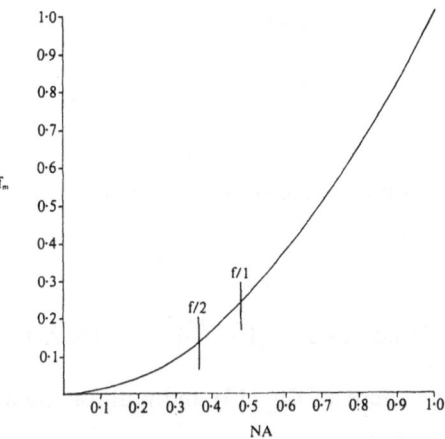

Fig. 8 Variation of light-collecting efficiency (f_m) of an optical fibre from a Lambertian source plotted against numerical aperture (N.A.). Theoretical efficiencies of perfect lenses working at 1:1 conjugates are indicated on this graph for two f/numbers.

C. The Effect of Angled End-Faces

Some applications demand that the end-faces of the bundle of optical fibres be curved. In such bundles, if there is a very large number of fibres, the end-faces of each fibre can be represented by planes at an angle to the cylindrical axis, these planes being tangential to the desired curve. To calculate the effect of this, consider the propagation of a meridional ray through a fibre in which the input end-face is inclined at an angle α to the axis and the normal to the end-face lies in the meridional plane.

Figure 9 shows a ray being transmitted by such a fibre. In this diagram, in addition to the normal symbols, β and β' are introduced to represent the incident and refracted angle respectively.

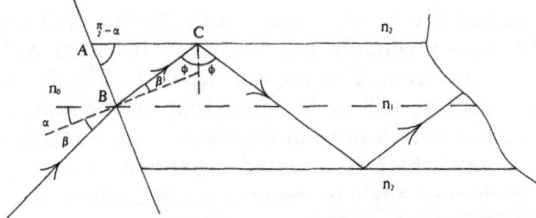

Fig. 9 A light ray being accepted by an optical fibre with an angled end-face.

From $\triangle ABC$, it is obvious that:

$$\frac{\pi}{2} - \alpha = \beta' + \phi$$

Therefore

$$-\beta' = \phi - (\frac{\pi}{2} - \alpha)$$

i.e.

$$\sin \beta' = \cos \alpha \cos \phi - \sin \alpha \sin \phi \qquad (17)$$

At the reflection limit:

$$\sin \phi = n_2/n_1$$

And $\cos \phi = (1 - \frac{n_2^2}{n_1^2})^{\frac{1}{2}} = \frac{1}{n_1}$ (N.A.)

By substituting these in Eq. (17) and multiplying across by n_1, we obtain:

$$n_1 \sin \beta' = (\text{N.A.}) \cos \alpha - n_2 \sin \alpha \qquad (18)$$

From Snell's law:

$$n_1 \sin \beta' = n_0 \sin \beta$$

Substituting this in Eq. (18) gives:

$$n_0 \sin \beta = \text{N.A.} \cos \alpha - n_2 \sin \alpha \qquad (19)$$

If the ray is oriented in the opposite sense with respect to the normal to the end-face, it can be shown that the reflection limit is reached when:

$$n_0 \sin \beta = \text{N.A.} \cos \alpha + n_2 \sin \alpha \qquad (20)$$

Since the above calculation was done at the critical angle, Eqs. (19) and (20) are the equivalents, for an angled end-face, to Eq. (7).

In this calculation, the axial incidence angle θ is given by $\theta = \alpha \pm \beta$. For a normal end-face $\alpha = 0$ and $\beta = \theta$, which yields the expected result:

$$n_0 \sin \beta = \text{N.A.}$$

The two values for $n_0 \sin \beta$ arise from the fact that the reflection conditions for rays positively and negatively oriented with respect to the normal are different. For the ray shown in Fig. 8, the incidence angle at the boundary will be less than that for the same ray entering the same fibre but with a normal end-face. Thus for this ray Eq. (19) applies. For a ray which is oriented negatively at an angle β to the normal, the incidence angle is greater than in the former case, and for this, Eq. (20) applies.

If light is being collected from a Lambertian source, then the maximum efficiency occurs for an N.A. of greater than unity because of Eq.

(19). Maximum efficiency will be attained when the ray which is positively oriented at $\pi/2$ to the normal is just reflected, and this occurs when:

$$n_0 = \cos \alpha \, (\text{N.A.}) - n_2 \sin \alpha \qquad (21)$$

If $n_0 = 1.0$ Eq. (21) becomes:

$$\text{N.A.}_\alpha = \frac{1 + n_2 \sin \alpha}{\cos \alpha} \qquad (22)$$

Thus, for a fibre with an angled end-face to collect all the light from a Lambertian source, its N.A._α must at least equal that given by Eq. (22). Figure 10 gives the variation of N.A. with α for $n_2 = 1.5$

If the output face of the fibre is normal to the axis, then the output axial angle θ will be related to the reflection angle ϕ by Eq. (3). That is:

$$n_0 \sin \theta = n_1 \cos \phi$$

Substituting for ϕ in Eq. (17) gives:

$$\sin \beta' = \frac{n_0}{n_1} \sin \theta \cos \alpha - \left(1 - \frac{n_0^2 \sin^2 \theta}{n_1^2}\right)^{\frac{1}{2}} \sin \alpha \qquad (23)$$

Multiplying across by n_1 and applying Snell's law to the left-hand side yields:

$$n_0 \sin \beta = n_0 \sin \theta \cos \alpha - (n_1^2 - n_2^2 \sin^2 \theta)^{\frac{1}{2}} \sin \alpha \qquad (24)$$

The right-hand side of Eq. (24) is less than unity for $\sin \theta = 1.0$, i.e. $\theta = \pi/2$.

In other words, there exists a range of values of β for the ray shown

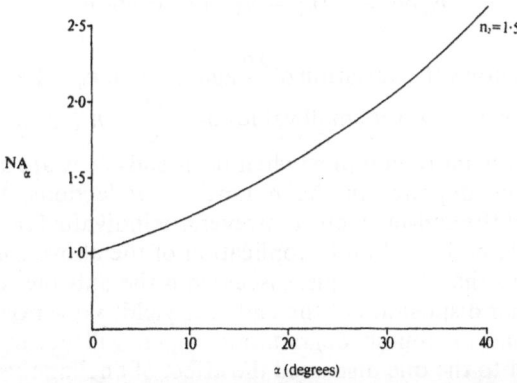

Fig. 10 Variation of minimum numerical aperture (N.A._α) for maximum efficiency of an optical fibre with angled input face plotted against input face inclination (α).

for which internal reflection will occur at a normal end-face. This range is given by the inequality:

$$n_0 \sin \beta > n_0 \cos \alpha - (n_1^2 - n_2^2)^{\frac{1}{2}} \sin \alpha \qquad (25)$$

The right-hand side of this inequality (25) is obtained by substituting $\sin \theta = 1.0$ in the right-hand side of Eq. (24). So, although a high N.A. can enable a fibre with an angled end-face to collect all the light from a Lambertian source, the efficiency of the fibre as a whole will be less than unity owing to internal reflection at the output face.

The effect of an angled end-face is also important when considering the degradation of performance due to manufacturing tolerances. As will be seen later, it is impossible to finish some fibre bundles so that all the fibres have their end-faces normal to the axis. The effect of this is to degrade the maintenance of axial angle, but in practice, this variation from normality will be small, and one can set $\sin \alpha = \alpha$ and $\cos \alpha = 1.0$ in equation (24) yielding:

$$n_0 \sin \beta = n_0 \sin \theta - \alpha (n_1^2 - n_0^2 \sin^2 \theta)^{\frac{1}{2}} \qquad (26)$$

The corresponding equation for an oppositely directed ray is:

$$n_0 \sin \beta = n_0 \sin \theta + \alpha (n_1^2 - n_0^2 \sin^2 \theta)^{\frac{1}{2}} \qquad (27)$$

If α is small, one can write $\beta = \theta \pm \delta\theta$ where $\delta\theta \ll \theta$.
Thus:

$$n_0(\sin \theta \pm \cos \theta \, \delta\theta) = n_0 \sin \theta \pm \alpha (n_1^2 - n_0^2 \sin^2 \theta)^{\frac{1}{2}}$$

by combining Eqs. (26) and (27).
Therefore:

$$n_0 \, \delta\theta = \alpha (n_1^2 - n_0^2 \sin^2 \theta)^{\frac{1}{2}}/\cos \theta \qquad (28)$$

Figure 11 shows the variation of $\dfrac{\delta\theta}{\alpha}$ against θ for $n_0 = 1$ and $n_1 = 1.62$. As would be expected for small values of θ, $\dfrac{\delta\theta}{\alpha} \sim n_1$, as in a thin prism.

The situation is more complex when both end-faces are angled since the conditions depend on the number of reflections, the relative orientation of the end-faces etc. However, for individual rays, the paths can be calculated by a double application of the above formulae.

It is obvious that the case just discussed is the only meridional case, since any other disposition of the end-face yields skew rays. Also, this case shows the maximum degradation which will occur, and in the plane normal to the one discussed the effect of angling the end-face is a minimum.

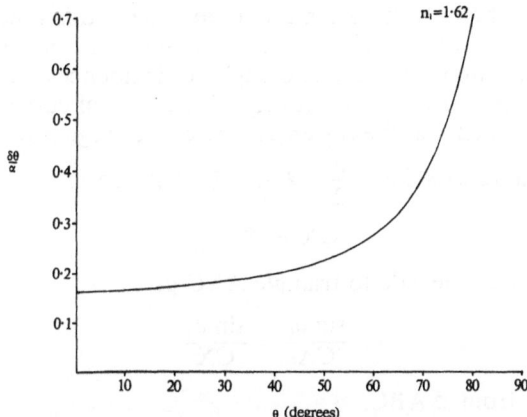

Fig. 11   Variation of fractional increase in output axial angle ($\delta\theta$) for an optical fibre with angled input face, plotted against input axial angle (θ); $\alpha \ll 1$.

D. The Effect of Fibre Curvature

One of the important properties of systems involving optical fibres is that they can follow a curved path or be flexed, so it is instructive to consider the effect of a bend on the passage of a light ray through the fibre.[4] Figure 12 shows a fibre of diameter d and bent to a radius R, with a meridional ray passing through it; it will be noted that the plane

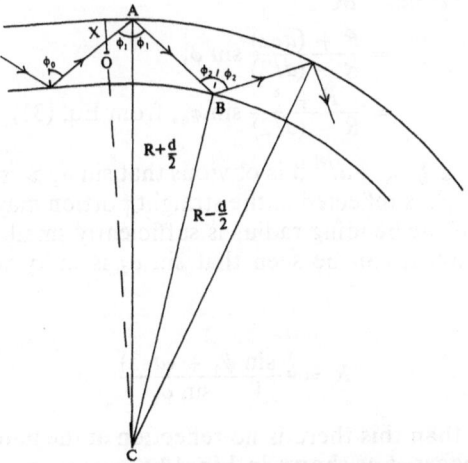

Fig. 12   Passage of typical meridional ray through a curved fibre.

illustrated is the only meridional one possible in this situation. In the straight portion, to the left of CX, the angle of incidence is ϕ_0 and in the curved portion there are two angles of incidence ϕ_1 and ϕ_2 for the outer and inner surfaces respectively. The centre of curvature is C and it will be assumed that the ray enters the curved region at X, a distance ξ from the fibre axis, i.e. $-\dfrac{d}{2} < \xi < +\dfrac{d}{2}$. Therefore:

$$CX = R + \xi$$

Applying the sine rule to triangle AXC gives:

$$\frac{\sin \phi_0}{CA} = \frac{\sin \phi_1}{CX} \tag{29}$$

Similarly, from $\triangle ABC$:

$$\frac{\sin \phi_1}{BC} = \frac{\sin \phi_2}{AC} \tag{30}$$

From Eq. (29) it can be shown that:

$$\sin \phi_1 = \frac{CX}{CA} \sin \phi_0$$

$$= \frac{R + \xi}{R + (d/2)} \sin \phi_0 \tag{31}$$

And from Eq. (30):

$$\sin \phi_2 = \frac{AC}{BC} \sin \phi_1$$

$$= \frac{R + (d/2)}{R - (d/2)} \sin \phi_1$$

$$= \frac{R + \xi}{R - (d/2)} \sin \phi_0, \text{ from Eq. (31).} \tag{32}$$

Since $-d/2 \ll \xi \ll +d/2$ it is obvious that $\sin \phi_2 \gg \sin \phi_0 \gg \sin \phi_1$. Thus a ray which is reflected in the straight portion may escape at the outer surface if the bending radius is sufficiently small.

From Eq. (32) it can be seen that $\sin \phi_2$ is unity when $\sin \phi_0 = \dfrac{R - (d/2)}{R + \xi}$

or

$$R = \frac{\xi \sin \phi_0 + (d/2)}{1 - \sin \phi_0} \tag{33}$$

For R smaller than this there is no reflection at the inner surface, the ray being propagated as shown in Fig. 13.

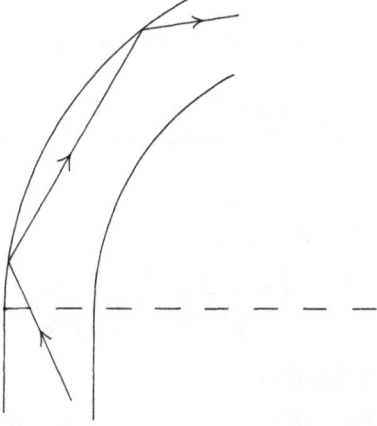

Fig. 13 Illustration of extreme meridional case where ray does not strike inner interface.

The extreme case for the ray in Eq. (31) is when $\xi = -d/2$ i.e. $\sin \phi_1 = \dfrac{R - (d/2)}{R + (d/2)} \sin \phi_0$.

This ray will be only just reflected when:

$$\sin \phi_1 = n_2/n_1 \tag{34}$$

By repeating the calculations done at the beginning of this section, a value for the maximum acceptance angle is obtained, given by:

$$\sin \theta_M = \left[n_1^2 - \left(\frac{R + (d/2)}{R - (d/2)}\right)^2 n_2^2 \right]^{\frac{1}{2}} \tag{35}$$

$$= \left[n_1^2 - \left(1 + \frac{d}{R}\right) n_2^2 \right]^{\frac{1}{2}} \text{ when } R \gg d/2$$

The effect of bending can be seen to have a rather drastic effect on the meridional rays; however, these form a much smaller proportion of the flux passing through the fibre than they do with a straight fibre, since there is only one meridional plane.

The path length in the curved fibre between reflections is AB and by applying the sine rule to the triangle ABC the following expression is obtained:

$$\frac{AB}{\sin \angle ACB} = \frac{BC}{\sin \phi_1}$$

or

$$AB = [R - (d/2)] \sin \angle ACB/\sin \phi_1 \qquad (36)$$

This corresponds to a length of fibre given by the arc defined by the radii CA and CB, which is $R \times \angle ACB$. Therefore, the path length per unit length is given by:

$$l'_m = \frac{AB}{R \times \angle ACB}$$

$$= \frac{\left(R - \dfrac{d}{2}\right)}{R} \cdot \frac{\sin \angle ACB}{\angle ACB} \cdot \frac{1}{\sin \phi_1} \qquad (37)$$

Substituting from Eq. (31):

$$l'_m = \frac{\left(R - \dfrac{d}{2}\right)}{R} \cdot \frac{\sin \angle ACB}{\angle ACB} \cdot \frac{R + \dfrac{d}{2}}{R + \xi} \cdot \frac{1}{\sin \phi_0} \qquad (38)$$

Re-arranging this and using the fact that the path length per unit length in the unbent fibre, l_m, is given by $1/\cos \theta \ (= 1/\sin \phi_0)$:

$$l'_m = \frac{\left(R^2 - \dfrac{d^2}{4}\right)}{R^2 + R\xi} \cdot \frac{\sin \angle ACB}{\angle ACB} \cdot l_m \qquad (39)$$

Since, in Eq. (39) the first term on the right-hand side is less than unity and $\sin \alpha < \alpha$, it is obvious that $l'_m < l_m$. That is, the path length of a ray in a bent fibre is less than the path length of a corresponding ray in a straight fibre. Also, the number of reflections per unit length is given by:

$$\eta'_m = \frac{1}{R \times \angle ACB} \qquad (40)$$

By the sine rule in $\triangle$ ABC, it can be shown that:

$$R = \frac{d}{2} \cdot \frac{(\sin \phi_1 + \sin \phi_2)}{(\sin \phi_2 - \sin \phi_1)} \qquad (41)$$

And $\angle ACB = \phi_2 - \phi_1$

Therefore $R \times \angle ACB = \dfrac{d}{2} \cdot \dfrac{(\sin \phi_2 + \sin \phi_1)}{(\sin \phi_2 - \sin \phi_1)} \cdot (\phi_2 - \phi_1) \qquad (42)$

If $\delta = \phi_2 - \phi_1$ then

$$\sin \phi_2 = \sin \phi_1 \cos \delta + \cos \phi_1 \sin \delta$$

Substituting this in Eq. (42) yields:

$$R \times \angle ACB = \frac{d}{2} \cdot \frac{(\sin \phi_1 \cos \delta + \sin \phi_1 + \cos \phi_1 \sin \delta)\delta}{\sin \phi_1 \cos \delta - \sin \phi_1 + \cos \phi_1 \sin \delta} \quad (43)$$

In most applications, $R \gg r$ and therefore δ is small, so that:

$$\cos \delta = 1.0; \sin \delta = \delta$$

Therefore:

$$R \times \angle ACB = \frac{d}{2}(2 \tan \phi_1 + \delta) \quad (44)$$

If Eq. (44) is substituted in (40) the expression becomes:

$$\eta'_m = \frac{1}{d \tan \phi_1 + \dfrac{\delta d}{2}} \quad (45)$$

By comparison with Eq. (11) it can be seen that the number of reflections per unit length is less than that in a straight fibre where the angle of reflection is ϕ_1. However, $\tan \phi_1 > \tan \phi_0$ so it is difficult to compare the above with the original ray in the unbent fibre.

E. The Effects of Varying Diameter

It is essential to know the effects upon propagation of variations in diameter, which are inevitable in a manufactured fibre. To simplify the mathematics, the discussion will be confined to the case of a rectilinear variation in diameter with length, i.e. a straight conical fibre. Figure 14 shows the path of a ray through such a section.

The section is defined by the extreme diameters and the conical semi-angle α. The meridional plane is any plane containing the axis, and it is obvious from this that successive normals are disposed at a relative angle of -2α. Thus, for the direction of propagation shown, the angles of incidence decrease by 2α at each reflection. For certain rays, the

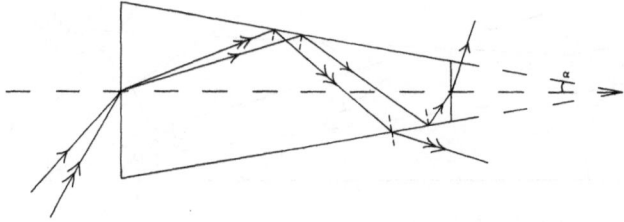

Fig. 14 Passage of meridional rays through a conical optical fibre.

angle of incidence will become less than the critical angle for reflection, and the ray will escape from the fibre, while for rays that do not escape it is obvious that the exit angle θ_2 will be greater than the original axial angle θ_1. For propagation in the reverse direction, the incidence angle increases through the fibre, and therefore no rays escape, in fact a ray can be captured by the fibre if it enters through the sheath at the appropriate angle. This can be seen in Fig. 14 by reversing the direction of the rays, when it will also be seen that the exit angle is less than the input angle.

For a tapered fibre of circular cross-section, the meridional plane is any plane through the axis and the passage of a meridional ray can be represented as shown in Fig. 15. In this, the path is represented as a straight line traversing images of the fibre reflected through the reflection planes, and the actual path of the ray is shown as a dotted line.

The section is defined by radii r_1 and r_2 $(r_1 > r_2)$ and semi-angle α. The ray enters the fibre at a point X which is a distance δ_1 from the axis $(-r_1 < \delta_1 < +r_1)$, with an input angle θ_1, and refracted angle θ_1'. The ray leaves the fibre at a point Y, a distance δ_2 from the axis $(-r_2 < \delta_2 < +r_2)$ and with output angle θ_2, corresponding to an internal angle of θ_2'.

If the sine rule is applied to $\triangle$ OXY it is found that:

$$\frac{\sin \angle \text{OXY}}{\text{OY}} = \frac{\sin \angle \text{XYO}}{\text{OX}} \tag{46}$$

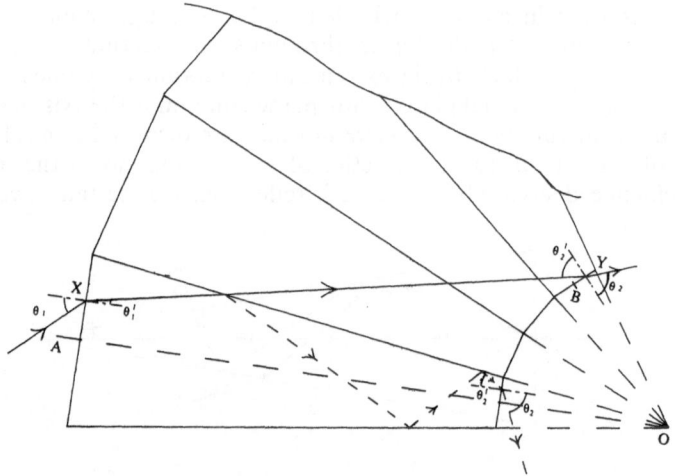

Fig. 15 Representation of the passage of a typical meridional ray through a conical optical fibre.

also
$$\left. \begin{array}{l} \angle OXY = \theta'_1 + \angle AOX \\ \angle XYO = \dfrac{\pi}{2} - \theta'_2 + \angle BYO \end{array} \right\} \quad (47)$$

Substituting from Eq. (46) in Eq. (47) yields:
$$\frac{\sin(\theta'_1 + \angle AOX)}{OY} = \frac{\sin(\pi/2 - \theta'_2 + \angle BYO)}{OX}$$

Therefore:
$$\frac{\sin\theta'_1 \cos\angle AOX + \cos\theta'_1 \sin\angle AOX}{OY} =$$

$$\frac{\cos\theta'_2 \cos\angle BYO + \sin\theta'_2 \sin\angle BYO}{OX} \quad (48)$$

Now, $\cos\angle AOX = \dfrac{OA}{OX} = \dfrac{r_1 \cot\alpha}{OX}$; $\cos\angle BYO = \dfrac{BY}{OY} = \dfrac{\delta_2}{OY}$

$\sin\angle AOX = \dfrac{\delta_1}{OX}$; $\sin\angle BYO = \dfrac{OB}{OY} = \dfrac{r_2 \cot\alpha}{OY}$

If these values are substituted in Eq. (48) and cross-multiplied by OX and OY, the folliwing expression is obtained:
$$r_1 \sin\theta'_1 \cot\alpha + \delta_1 \cos\theta'_1 = \delta_2 \cos\theta'_2 + r_2 \sin\theta'_2 \cot\alpha \quad (49)$$

Re-arranging Eq. (49) and multiplying both sides by $\tan\alpha$ gives:
$$r_1 \sin\theta'_1 + \delta_1 \tan\alpha \cos\theta'_1 = r_2 \sin\theta'_2 + \delta_2 \tan\alpha \cos\delta'_2 \quad (50)$$

Now $\delta\cos\theta < r$, so that if $\alpha \ll \theta'_1$, an approximation to Eq. (50) can be obtained, which is:
$$r_1 \sin\theta'_1 = r_2 \sin\theta'_2 \quad (51)$$

To transform to exterior angles multiply across by n_1 and apply Snell's law, that is:
$$r_1 \sin\theta_1 = r_2 \sin\theta_2 \quad (52)$$

Thus, for small angles the system obeys the sine condition. It is obvious from this that one cannot increase the brightness of a light beam by passing it through a conical fibre, although one can increase the flux per unit area emerging from the output face.

Clearly, the maximum possible output axial angle will be determined by the N.A. of the fibre, i.e.
$$\sin\theta_2 (\max) = \frac{1}{n_0}(n_1^2 - n_2^2)^{\frac{1}{2}}$$

Substituting in Eq. (52) one finds that the maximum value of θ_1 for which propagation is possible is given by:

$$\sin \theta_1 \, (\text{max}) = \frac{r_2}{r_1} \times \frac{1}{n_0} (n_1^2 - n_2^2)^{\frac{1}{2}} \qquad (53)$$

Figure 16 shows the variation of θ_1 (max) with r_2/r_1 for a number of values of N.A. If $\delta_1 = \delta_2 = 0$ it can be shown that the path length in the section is given by:

$$X\,Y = r_2 \cot \alpha \, \frac{\sin (\theta_2' - \theta_1')}{\sin \theta_1'}$$

The length of the section is $(r_1 - r_2) \cot \alpha$ so that the path length per unit length is given by:

$$l''_m = \frac{r_2}{r_1 - r_2} \cdot \frac{\sin (\theta_2' - \theta_1')}{\sin \theta_1'} \qquad (54)$$

Further, the number of reflections is given by $\dfrac{\theta_2' - \theta_1'}{2\alpha}$ so that the number of reflections per unit length is given by:

$$\eta''_m = \frac{\theta_2' - \theta_1'}{2\alpha \, (r_1 - r_2) \cot \alpha} = \frac{\theta_2' - \theta_1'}{2 \, (r_1 - r_2)}, \quad \text{when } \alpha \ll 1 \qquad (55)$$

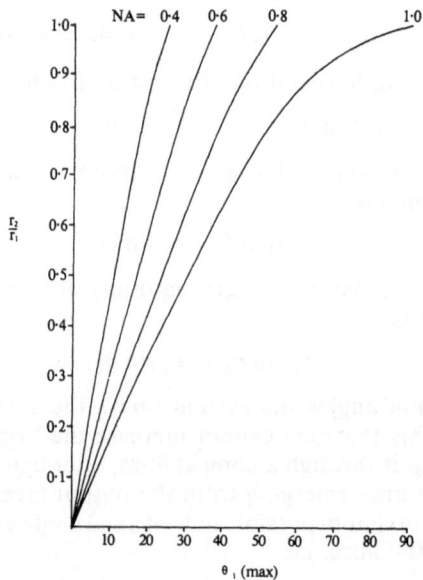

Fig. 16 Variation of maximum input angle ($\theta_{1(max)}$) for a meridional ray in a conical fibre plotted against the ratio of the defining radii (r_2/r_1).

Because of the flux density increase that can be obtained by passing a light beam through a conical fibre, the tapered fibre is used as an optical component in its own right.

If the above is being used to consider the effects on propagation of small variation in diameter, one can write $r_2 = r_1 - \Delta r_1$, where $\Delta r_1 \ll r_1$, and $\theta'_2 = \theta'_1 + \Delta \theta'_1$ where $\Delta \theta'_1 \ll 1$.

Substituting in Eq. (51) yields:

$$r_1 \sin \theta'_1 = (r_1 - \Delta r_1) \sin (\theta'_1 + \Delta \theta'_1)$$
$$= r_1 \sin \theta'_1 \cos \Delta \theta'_1 + r_1 \cos \theta'_1 \sin \Delta \theta'_1 - \Delta r_1 \sin \theta'_1 \cos \Delta \theta'_1 - \Delta r_1 \cos \theta'_1$$
$$\times \sin \Delta \theta'_1 \qquad (56)$$

Since $\Delta \theta'_1 \ll 1$ and $\Delta r_1 \ll r_1$, one can write: $\cos \Delta \theta'_1 = 1.0, \sin \Delta \theta'_1 = \Delta \theta'_1$ and $\Delta r_1 \sin \theta = 0$ which gives:

$$r_1 \cos \theta'_1 \times \Delta \theta'_1 = \Delta r_1 \sin \theta'_1 \qquad (57)$$

Also:

$$l''_m = \frac{r_1 - \Delta r_1}{\Delta r_1} \cdot \frac{\Delta \theta'_1}{\sin \theta'_1} = \frac{r_1 - \Delta r_1}{\sin \theta'_1} \cdot \frac{\Delta \theta'_1}{\Delta r_1} \qquad (58)$$

Substituting for $\Delta \theta'_1 / \Delta r_1$ from Eq. (57) gives:

$$l''_m = \frac{r_1 - \Delta r_1}{r_1} \cdot \frac{1}{\cos \theta'_1} \qquad (59)$$

Similarly:

$$\eta''_m = \frac{\Delta \theta'_1}{2 \Delta r_1} = \frac{\tan \theta'_1}{2 r_1} \qquad (60)$$

When $r_1 = r_2$, it will be seen that Eqs. (59) and (60) become identical to Eqs. (10) and (11) for a straight fibre. Also for small variations in diameter, the changes in unit path length $(\Delta l''_m)$ and unit number of reflections $(\Delta \eta''_m)$ are given by:

$$\left. \begin{array}{c} \dfrac{\Delta l''_m}{l''_m} = - \dfrac{\Delta r}{r} \\[2ex] \dfrac{\Delta \eta''_m}{\eta''_m} = 0 \end{array} \right\} \qquad (61)$$

III. SKEW RAY ANALYSIS

A. Propagation of Skew Rays

In the previous section the propagation of meridional rays through fibres has been dealt with at reasonable length, which has been possible owing to the relative simplicity of the mathematical description. In this section skew rays will be considered, i.e. rays whose passage through the fibre is not confined to a single plane. Because of the increased complexity of the mathematics only the propagation of these rays along a straight cylindrical fibre will be discussed.[1,3] In Fig. 17 a typical skew ray XY is depicted, where X and Y are the points of reflection at the cylinder wall. P is the foot of the perpendicular from Y to a plane PCX normal to the fibre axis, where C is the intersection of the axis with this plane. The azimuthal angle, γ, is defined as the angle between PX and XC. Since PY is parallel to the axis, θ' is the internal axial angle, as used in previous sections. Thus γ and $(\frac{\pi}{2} - \theta')$ define the orientation of the ray XY with respect to the normal, CX, at X. Since these angles are defined in perpendicular planes, we can write for the angle of reflection, ϕ:

$$\cos \phi = \cos \angle CXY = \sin \theta' \cos \gamma \qquad (62)$$

Figure 18 shows the projections of a number of reflections of such a ray on the plane CPX. Thus for any skew ray, the angle γ is maintained during propagation, and by considering a plane parallel to the axis in Fig. 17 at the point Y, it is obvious that the angle θ' is also maintained. Since Eq. (62) is applicable to any ray at any point in the fibre, it will be seen that the reflection angle ϕ is also maintained for each reflection. By multiplying Eq. (62) by n_1 and applying Snell's law, the condition for propagation can be derived in a manner similar to that of section II A. The acceptance condition for skew rays is found to be:

$$n_0 \sin \theta \cos \gamma < (n_1^2 - n_2^2)^{\frac{1}{2}} \qquad (63)$$

Alternatively, the maximum acceptance angle, Θ_M is given by:

$$\sin \Theta_M = \frac{1}{\cos \gamma} \cdot \frac{1}{n_0} (n_1^2 - n_2^2)^{\frac{1}{2}} = \frac{\sin \theta_M}{\cos \gamma} \qquad (64)$$

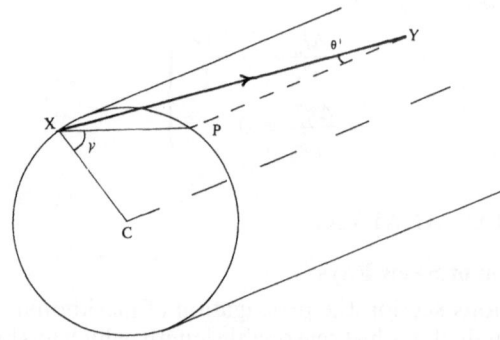

Fig. 17   Path of a typical skew ray between reflections.

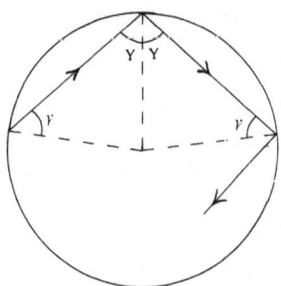

Fig. 18   Projection of the path of a typical skew ray on a plane normal to the fibre axis.

Thus skew rays will be accepted at larger axial angles than meridional rays, where $\cos \gamma = 1.0$, so that the meridional N.A. is a minimum figure; indeed, for any N.A. Θ_M can be equal to 90° for values of γ greater than $\frac{\pi}{2} - \theta_M$.

If a beam of parallel light is incident on the face of an optical fibre at an angle θ to the axis, then the angle γ divides the fibre face into three areas as shown in Fig. 19, separated by lines parallel to the projection of the beam direction on the face. In the areas which are shaded, the rays have a value of γ which is in excess of γ_1, say. Now if the angle θ is in excess of θ_M then the light incident on the shaded areas will be accepted if γ_1 is defined by $\cos \gamma_1 = \dfrac{\sin \theta_M}{\sin \theta}$.

By reversing this argument, one can see that if an optical fibre, with N.A. < 1.0, is illuminated from a diffuse source, and the output is

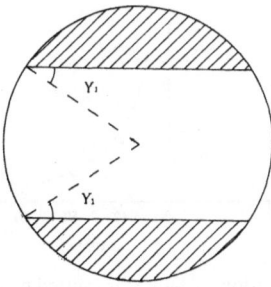

Fig. 19   Illustration of zones on fibre cross-section which have a value of azimuthal angle, γ, in excess of γ_1.

observed at an angle θ, greater than θ_M, to the axis, then light will be seen to emerge only from the areas where cos γ is greater than $\dfrac{\sin \theta_M}{\sin \theta}$. Thus a black band[5] will be seen across the fibre whose width increases as the viewing angle is increased. Even at 90° to the axis there will be light emitted in the areas where cos γ is greater than $\sin \theta_M$. The width of the band is given by t, where $t = d \sin \gamma = d\left(1 - \dfrac{\sin^2 \theta_M}{\sin^2 \theta}\right)^{\frac{1}{2}}$, where d is the fibre diameter. It can be shown that the area of the shaded regions is given by A_s, where $A_s = \dfrac{d^2}{4}(\pi - 2\gamma - \sin 2\gamma)$ and that the fractional area this represents, A_f, is:

$$A_f = 1 - \frac{(2\gamma + \sin 2\gamma)}{\pi} \qquad (65)$$

Figure 20 shows the amount of light transmitted by an optical fibre plotted against axial angle for an N.A. of 0.5 where the light transmitted at angles in excess of 30° has been calculated using Eq. (65). The crosses indicate experimental points obtained for an actual fibre.

It will be obvious by studying Fig. 18 that skew rays travel down the fibre in helical fashion and, when observing an optical fibre at angles greater than the acceptance angle, that the light emerging from one

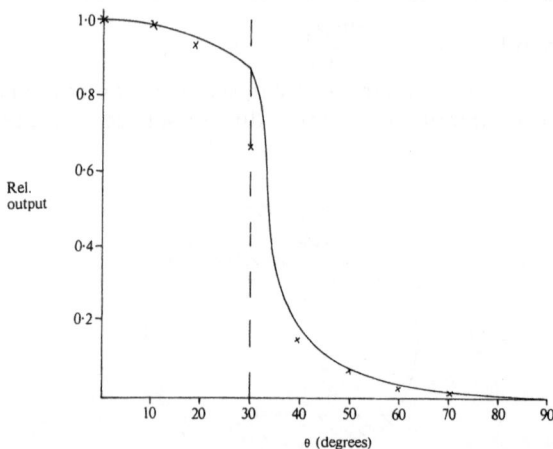

Fig. 20 Variation of the fraction of light transmitted by an optical fibre (N.A. = 0.5) from a Lambertian source, assuming zero absorption and reflection losses. The crosses are normalised experimental values.

side of the black band has travelled through the fibre in a helix of the opposite sense to the light coming from the other side.

By applying the same arguments as were used in section IIA, it can be shown that the unit path length (l_s) and number of reflections (η_s) are given by:

$$l_s = \frac{1}{\cos \theta'} = l_m \tag{66}$$

And

$$\eta_s = \frac{\tan \theta'}{d \cos \gamma} = \frac{\eta_m}{\cos \gamma} \tag{67}$$

It can be seen from Eq. (66) that the path length in an optical fibre is dependent solely on the axial angle. This is an important fact when considering the use of optical fibres with coherent light. It will also be seen, from Eq. (67), that the number of reflections increases significantly as $\gamma \longrightarrow \pi/2$ and, at the limit, the path of the ray will be a true helix tangential to the reflecting surface.

It is obvious from Fig. 18 that the direction of any ray changes by 2γ at each reflection in the plane normal to the axis. Because γ varies uniformly from $0°$ to $90°$ across the face of the fibre, the knowledge of the incident direction, in this plane, of any light ray will soon be lost after a number of reflections. On the other hand, the incident axial angle will be maintained in the output beam. Thus a beam of parallel light incident at the fibre face at an axial angle θ will emerge as a hollow cone of light of semi-angle θ.

B. Light-Gathering Power of an Optical Fibre

The light-gathering power of an optical fibre will obviously be greater than that calculated in section IIB when the effects of skew rays are included, since these will be accepted at higher input angles than meridional rays. Potter[6] has calculated the total light-gathering power, P, of an optical fibre which can be represented as follows:

$$P = n_0^2 - \frac{2}{\pi} \left[(n_1^2 - n_2^2)^{\frac{1}{2}} (n_0^2 - n_1^2 + n_2^2)^{\frac{1}{2}} + \right.$$

$$\left. (n_0^2 - 2(n_1^2 - n_2^2) \cos^{-1} \frac{(n_1^2 - n_2^2)^{\frac{1}{2}}}{n_0} \right] \tag{68}$$

where n_0, n_1 and n_2 have their usual meanings.

By substituting in Eq. (68) $n_0 = 1.0$ and $(n_1^2 - n_2^2) = (\text{N.A.})^2$ one obtains:

$$P = 1 - \frac{2}{\pi} \{ \text{N.A.} (1 - (\text{N.A.})^2)^{\frac{1}{2}} + (1 - 2(\text{N.A.})^2) \cos^{-1} (\text{N.A.}) \} \tag{69}$$

Dividing Eq. (69) by the meridional light-gathering powew (N.A.)²
we obtain:

$$\frac{P}{(N.A.)^2} = \frac{1}{(N.A.)^2} - \frac{2}{\pi}\left\{\left(\frac{1}{(N.A.)^2} - 1\right)^{\frac{1}{2}} + \left(\frac{1}{(N.A.)^2} - 2\right)\cos^{-1}(N.A.)\right\}$$

$$(70)$$

Figure 21 shows the variation of the ratio of total light-gathering
power to meridional light-gathering power with N.A., from which it
will be seen that, as the N.A. tends to unity, the ratio also tends to
unity as would be expected, since there is then no contribution by skew
rays for angles in excess of the meridional acceptance angle. Also, the
ratio increases as the N.A. decreases, which is again understandable
since the smaller the N.A. then the greater is the range of angles in
which the skew rays can contribute in excess of the meridional situation. However, it is also true to say that the N.A. does enable a measure
of the light-gathering power to be ascertained by means of a simple
calculation. It is for this reason that the use of the N.A. as a description
of the optical properties of fibres has been universally adopted.

It was shown in the previous section that the meridional case is the
same for any of the cross-sections normally used. However, this is not
the case for skew rays since, although the relations (65) and (64) hold,
the variation of γ across the fibre face is different. Figure 22 depicts the
projection of the path of a skew ray on the plane normal to the axis for
both square and hexagonal cross-sections. In Fig. 22(a), a square cross-

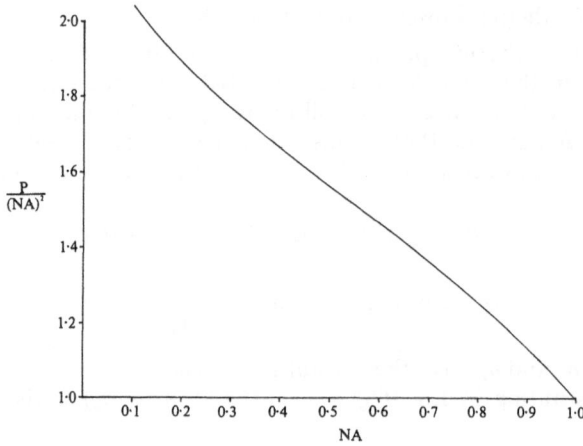

Fig. 21   Variation of the ratio P/(N.A.)² plotted against numerical aperture (N.A.) for
a circular optical fibre.

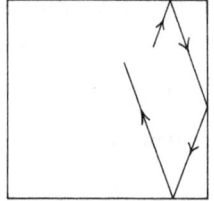 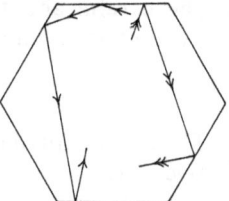

Fig. 22   Projection of the paths of skew rays in optical fibres of square and hexagonal cross-sections on to a plane normal to the fibre axis.

section is shown, and it is obvious that for any ray there are two values of γ which correspond to opposite pairs of sides, and which differ by $\pi/2$. In Eqs. (63) and (64) we must therefore use the smaller value. Further, for any ray the values of γ are determined by the direction of the incident ray, relative to the sides of the square, and for a parallel beam the values of γ are the same for all rays in the beam.

Since we must use the smaller value in our calculations, the maximum value of γ is $\pi/4$. From Eq. (64) we see that the maximum acceptance angle for skew rays is given by:

$$\sin \Theta_M = \sqrt{2} \, \sin \theta_M$$

Similarly, in the hexagonal case, Fig. 22(b), there are two values of γ which differ, in this case, by $2\pi/3$. For hexagonal fibres, the maximum acceptance angle is given by:

$$\sin \Theta_M = 2 \sin \theta_M$$

Therefore, if a fibre of either cross-section is illuminated from a diffuse source, then the cone of light emerging from the other end will not have a constant semi-angle, if the N.A. is sufficiently small. In fact, the output beam will have a maximum semi-angle when measured in a plane parallel to the diagonals, and a minimum when measured in a plane parallel to the sides. Further, there is no "black-band" effect with square fibres, since γ is constant for a particular direction.

At angles above the meridional acceptance angle, light is only accepted if the ray, incident at an axial angle θ, has a value of γ greater than $\cos^{-1}\left(\dfrac{\text{N.A.}}{\sin \theta}\right)$. Referring back to Fig. 7, we see that this means that the elemental flux collected by the square fibre, at these angles, is given by:

$$dF_\gamma = (2\pi - 8\gamma) \, I_0 \sin \theta \, \cos \theta \, d\theta$$

$$= 2\pi \, I_0 \sin \theta \, \cos \theta \, d\theta - 8 \, I_0 \sin \theta \, \cos \theta \, \cos^{-1}\left(\frac{\text{N.A.}}{\sin \theta}\right) d\theta \quad (71)$$

One must be careful about the choice of limits when integrating Eq. (71). Since the maximum value of y is $\pi/4$ then the upper limit will be $\sin^{-1}\sqrt{2}$ (N.A.) if $\sin\theta_M < \dfrac{1}{\sqrt{2}}$, and if $\sin\theta_M > \dfrac{1}{\sqrt{2}}$ then the upper limit is $\pi/2$. The lower limit of integration will be θ_M; thus the total light collected is given by:

$$F_s = \int_0^{\theta_M} dF_m + \int_{\theta_M}^{A} dF_y$$

where dF_m is obtained from Eq. (14) and

$$A = \sin^{-1}\sqrt{2}\,(\text{N.A.}) \qquad \text{when } \sin\theta_M < \frac{1}{\sqrt{2}}$$

$$= \pi/2 \qquad\qquad\quad \text{when } \sin\theta_M > \frac{1}{\sqrt{2}}$$

Therefore,

$$F_s = \int_0^{A} 2\pi\,I_0 \sin\theta\cos\theta\,d\theta - 8\,I_0\int_{\theta_M}^{A}\sin\theta\,\cos^{-1}\!\left(\frac{\text{N.A.}}{\sin\theta}\right)\cos\theta\,d\theta$$

This can be integrated using the substitution $y = \cos^{-1}\!\left(\dfrac{\text{N.A.}}{\sin\theta}\right)$ in the right-hand term, giving

$$F_s = 4\,I_0\,(\text{N.A.})^2 \text{ when N.A.} < \frac{1}{\sqrt{2}}$$

$$= \pi I_0 - 4I_0\,(\cos^{-1}(\text{N.A.}) - \text{N.A.}\,(1 - (\text{N.A.})^2)^{\frac{1}{2}}) \text{ when N.A.} > \frac{1}{\sqrt{2}}$$

$$\tag{72}$$

Thus, the light-gathering power of a square fibre from a Lambertian source is given by:

$$P = \frac{4}{\pi}\,(\text{N.A.})^2 \text{ when N.A.} < \frac{1}{\sqrt{2}}$$

$$\left. = 1 - \frac{4}{\pi}\,(\cos^{-1}(\text{N.A.}) - \text{N.A.}\,(1 - (\text{N.A.})^2)^{\frac{1}{2}}) \text{ when N.A.} > \frac{1}{\sqrt{2}} \right\}$$

$$\tag{73}$$

Figure 23 shows the variation of $\dfrac{P}{(\text{N.A.})^2}$, as given by Eq. (73), with (N.A.).

The situation for a hexagonal fibre is complicated by the fact that

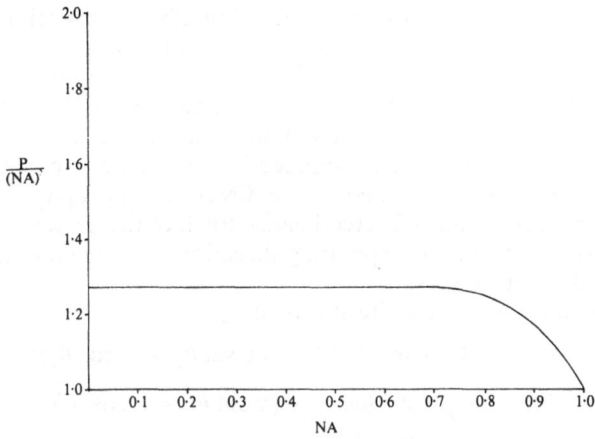

Fig. 23 Variation of the ratio P/(N.A.)² plotted against numerical aperture (N.A.) for a square optical fibre.

reflections can occur at each face or at alternate faces (see Fig. 22(b)). For the first, the maximum value of γ is $\pi/3$, and for the second, $\pi/6$. In the first case, the maximum acceptance angle is $\sin^{-1}$ (2N.A.) and is obtained when the light is incident in a plane normal to a pair of sides. However, this only applies to that portion of the beam which strikes one of the remaining sides, γ being zero otherwise. Thus a black band will be seen when the fibre is viewed in a direction normal to a side for an angle greater than $\sin^{-1}$ (N.A.). In the second case, the situation is really that of an equilateral triangle and the maximum acceptance angle is correspondingly lower and is given by $\sin^{-1} \left(\dfrac{2}{\sqrt{3}} \text{N.A.} \right)$. This is obtained when the fibre is viewed in a direction parallel to a diagonal, and, since the values of γ are constant, no black band is seen.

IV. PHYSICAL CONSIDERATION OF INTERNAL REFLECTION

Although geometric optics provides the condition for internal reflection to occur, it offers no explanation of the phenomenon or any information about the energy distribution at reflection. To gain further insight into this phenomenon, the results of electromagnetic theory must be applied to the case of a light ray impinging on a dielectric interface. The following discussion is based on the treatment outlined by Ditchburn.[7]

This treatment involves considering initially the situation where a
ray of light is incident at the dielectric boundary, and applying the
results of electromagnetic theory to obtain the fraction of the incident
energy contained in the reflected and refracted beams. This situation is
shown in Fig. 24, where the plane of incidence is the XZ plane. Three
new axes w_1, w_1' and w_2 are introduced in the plane of incidence which
are respectively perpendicular to the directions of propagation of the
incident, reflected and refracted beams. Each of these axes, along with
the OY axis and the corresponding direction of propagation, form a
right-handed set of axes.

The incident beam may be specified by:

$$E_{1w} = A_{1w} \exp i \{\omega t - \kappa_1(x \sin \theta_1 - z \cos \theta_1)\} \qquad (74)$$

$$E_{1y} = A_{1y} \exp i \{\omega t - \kappa_1(x \sin \theta_1 - z \cos \theta_1)\} \qquad (75)$$

The reflected and refracted beams can be similarly specified. By
solving the field equations using the boundary conditions, we obtain the
following equations, relating the reflected and refracted wave amplitudes to that of the incident wave:

$$A_{1w'} = A_{1w} \frac{n_2 \cos \theta_1 - n_1 \cos \theta_2}{n_2 \cos \theta_1 + n_1 \cos \theta_2} \qquad (76)$$

$$A_{1y'} = -A_{1y} \frac{n_2 \cos \theta_2 - n_1 \cos \theta_1}{n_2 \cos \theta_2 + n_1 \cos \theta_1} \qquad (77)$$

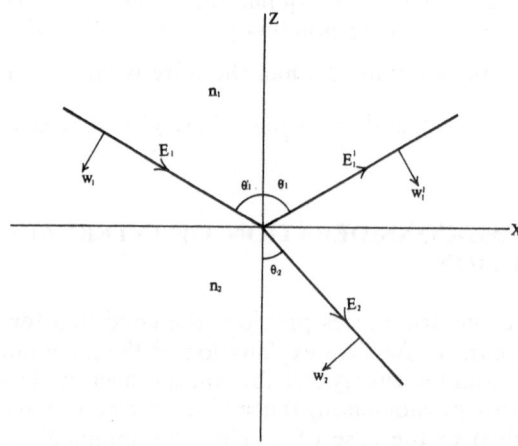

Fig. 24   Diagram showing the axes used to represent the reflection and refraction of a
light ray at a dielectric interface.

$$A_{2w} = A_{1w} \frac{2 \sin \theta_1 \cos \theta_1}{\sin (\theta_1 + \theta_2) \cos (\theta_1 - \theta_2)} \qquad (78)$$

$$A_{2y} = A_{1y} \frac{2 \sin \theta_2 \cos \theta_1}{\sin (\theta_1 + \theta_2)} \qquad (79)$$

If the angle of incidence is greater than the critical angle for internal reflection then, from Snell's law:

$$\sin \theta_2 = \frac{n_1}{n_2} \sin \theta_1$$

and

$$\cos \theta_2 = (1 - \left(\frac{n_1}{n_2}\right)^2 \sin^2 \theta_1)^{\frac{1}{2}} \qquad (80)$$

Now, $\sin \theta_1 > \frac{n_2}{n_1}$, so $\frac{n_1}{n_2} \sin \theta_1 > 1$. Therefore the cosine in Eq. (80) is imaginary, and can be written:

$$\cos \theta_2 = i \frac{n_1}{n_2} (\sin^2 \theta_1 - \left(\frac{n_2}{n_1}\right)^2)^{\frac{1}{2}} \qquad (81)$$

Substituting for Eq. (81) in Eq. (76) gives the following:

$$A_{1w'} = A_{1w} \frac{n_2 \cos \theta_1 - i \frac{n_1^2}{n_2} \left(\sin^2 \theta_1 - \left(\frac{n_2}{n_1}\right)^2\right)^{\frac{1}{2}}}{n_2 \cos \theta_1 + i \frac{n_1^2}{n_2} \left(\sin^2 \theta_1 - \left(\frac{n_2}{n_1}\right)^2\right)^{\frac{1}{2}}} \qquad (82)$$

Now any complex quantity $\frac{a - ib}{a + ib}$ can be written as $\exp (-2i\delta)$ where $\tan \delta = b/a$. Thus we can write:

$$A_{1w'} = A_{1w} \exp (-2i\delta_w)$$
$$E_{1w'} = A_{1w} \exp i \{\omega t - \kappa_1(x \sin \theta_1 + z \cos \theta_1) - 2\delta_w\} \qquad (83)$$

where $\tan \delta_w = \frac{n_1^2}{n_2^2} \frac{\left(\sin^2 \theta_1 - \frac{n_2^2}{n_1^2}\right)^{\frac{1}{2}}}{\cos \theta_1} \qquad (84)$

Similarly, it can be shown that:

$$E_{1y'} = A_{1y} \exp i \{\omega t - \kappa_1(x \sin \theta_1 + z \cos \theta_1) - 2\delta_y\} \qquad (85)$$

where $\tan \delta_y = \frac{\left(\sin^2 \theta_1 - \frac{n_2^2}{n_1^2}\right)^{\frac{1}{2}}}{\cos \theta_1} = \frac{n_2^2}{n_1^2} \tan \delta_w \qquad (86)$

Thus, the reflected wave has the same energy as the incident wave. The component of the electric vector in the plane of incidence is retarded by $2\delta_w$, and the perpendicular components by $2\delta_y$. Since the incident and reflected waves do not have a phase difference of π, there must be a disturbance in the second medium, in order to satisfy the boundary conditions. The components of the electric vector perpendicular to the plane of incidence is given by:

$$E_{2y} = A_{2y} \exp i \{\omega t - (\kappa_2 x \sin \theta_2 - \kappa_2 z \cos \theta_2)\} \qquad (87)$$

At the boundary:

$$
\begin{aligned}
A_{2y} &= A_{1y} + A_{1y'} \\
&= A_{1y}(1 + \exp(-2i\delta_y)) \\
&= 2 A_{1y} \exp(-i\delta_y) \cos \delta_y \qquad (88)
\end{aligned}
$$

By substituting from Eq. (88), $\sin \theta_2$ and $\cos \theta_2$ in Eq. (87) we obtain:

$$
\begin{aligned}
E_{2y} = 2A_{1y} \cos \delta_y \exp \{&- \kappa_1 z \left(\sin^2\theta_1 - \left(\frac{n_2}{n_1}\right)^2 z^{\frac{1}{2}}\right) \\
&+ i(\omega t - \kappa_1 \sin \theta_1 x - \delta_y)\} \qquad (89)
\end{aligned}
$$

The disturbance in the less dense medium is therefore periodic in x but not in z. Now the total disturbance in the denser medium is given by:

$$E_{1y} + E_{1y'} = 2A_{1y} \cos(\kappa_1 z \cos \theta_1 + \delta_y) \exp i(\omega t - \kappa_1 x \sin \theta_1 - \delta_y) \quad (90)$$

Thus, in the denser medium a stationary wave pattern is set up owing to interference between the incident and reflected beams, and in the less dense medium the amplitude falls exponentially, although the amplitudes are equal at the interface.

Therefore, electromagnetic theory predicts that internal reflection is total, i.e. no loss of energy, and that there is a disturbance in the second medium whose amplitude falls off exponentially with distance from the interface. This second conclusion is of great significance when dealing with optical fibres, since it means that complete isolation of the system from external influences is not possible.

Since the amplitudes on either side of the interface are equal, serious scattering of the beam can be caused by foreign particles in close proximity to the interface. This excludes the use of air (or any fluid) as the less dense medium, since the interface would not be protected against foreign particles settling on the boundary. Therefore, a solid material must be used for the less dense material and this must have a finite thickness so that the presence of foreign particles on its outer surface has a minimal effect on the guided beam. A reasonable thickness for minimal effect would be that at which the energy is 0.1% of the value at the interface. Figure 25 shows the variation of this value (measured in number of wavelengths) against incident angle for two

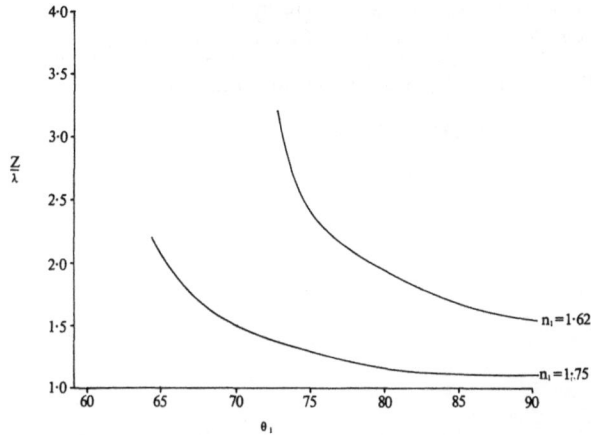

Fig. 25 Variation of the distance from the interface (Z/λ), measured in wavelengths, required to reduce the amplitude of the penetrated wave to 0.1% of its interfacial value. This is plotted against input axial angle for $n_1 = 1.62$ and 1.75 with $n_2 = 1.5$.

typical pairs of refractive indices. It can be seen that a thickness of 2λ to 3λ would be satisfactory for both situations. Similar results are obtained for the component of the electric vector in the plane of incidence.

These results agree with the observations of frustrated total reflections between two prisms reported by Quinke.[8]

The above discussion enables a full definition of an optical fibre to be made, as a cylinder of dielectric material (defined as the core), surrounded by a layer of a second solid dielectric (defined as the sheath) whose index is lower than that of the core, having a thickness of a few wavelengths. Although this step of settling for a solid sheath appears to restrict the range of N.A.s available, the net effect of this is advantageous and, in any case, N.A.s of greater than unity can in fact be achieved using a solid sheath. Conversely, the solid sheath allows the attainment of much lower N.A.s than would be possible if air were the less dense material.

REFERENCES

1. R. J. Potter, *J. Opt. Soc. Am.* **51**, 1079 (1961)
2. L. S. Allard, *Ind. Electronics* **2**, 273 (1964)
3. N. S. Kapany and D. F. Capallero, *J. Opt. Soc. Am.* **51**, 23 (1961)

4. N. S. Kapany *J. Opt. Soc. Am.* **47**, 413 (1957)
5. R. J. Potter, Ph.D Thesis, University of Rochester (1960)
6. R. J. Potter, E. Donath and R. Tynan, *J. Opt. Soc. Am.* **53**, 256 (1963)
7. R. W. Ditchburn, *Light*, Blackie and Son, Ltd., London and Glasgow (1952), p. 423
8. G. Quinke, *Pogg Ann.* **117**, 1117 (1863)

CHAPTER 3

The Optical Fibre

In the previous chapter an outline was given of the theoretical properties of a specialized optical system which has been called an optical fibre. From the results obtained we can now specify the form which a practical optical fibre must take.

Basically, the optical fibre is a cylinder of transparent dielectric material surrounded by a second dielectric. In order to obtain guidance, the refractive index of the cylinder material must be greater than that of the surrounding material and to obtain satisfactory optical isolation of the fibre, the surrounding material should have a thickness of one or two wavelengths of the light to be guided. In practical terms, this latter condition also requires the use of a solid surrounding material. A practical optical fibre would then consist of a cylindrical core of a glass or plastic sheathed by a thin layer of a second glass or plastic with a refractive index less than that of the core material.

In this chapter, the manufacture of this basic fibre will be outlined, and its properties discussed, bringing in the practical considerations of absorption and reflection losses, and other deviations of the practical fibre from the ideal case discussed in the previous chapter.

I. PRODUCTION OF OPTICAL FIBRES

In the field of fibre optics, the range of applications demands the use of optical fibres with sizes down to $10\,\mu$m, which clearly eliminates a direct manufacturing process. A technique must therefore be found which permits the component materials to be handled in practicable sizes, the desired size being produced during the manufacturing process. Such a technique was used by Soutter[1] who produced fine filaments of glass rod and tubing by mounting the rod or tube, on the end of which a weight was hung, vertically in a furnace which was then taken up to the softening temperature of the glass. When this temperature was achieved, the softened glass stock elongated under the action of the weight with a consequent reduction in stock diameter. Similar techniques can be found in the manufacture of fibre glass where rods or spheres of glass are heated to softening point, and fine filaments are drawn from these. A modified version of this principle forms the pre-

sent production method for the vast majority of optical fibres manufactured today.

The basic process is illustrated in Fig. 1. The stock of raw material is
mounted within a vertical muffle furnace which is held at a temperature
sufficient to cause the end of this stock to soften. The raw material is
drawn from the bottom of the furnace, and this elongation produces a
corresponding reduction in diameter; thus a cylinder of softened
material issues from the furnace and cools down to the ambient temperature. At a certain distance below the furnace the cylinder temperature will reach the setting point of the raw material and will solidify,
thus determining the final size produced. Now a liquid cylinder is unstable under the action of surface tension forces,[2] which act to break
up the cylinder into a series of spheres (this instability is readily observed in the action of a jet of water). However, if the liquid in the
drawn cylinder has a high viscosity this effect can be slowed down
considerably so that the cylinder is still intact when the solidification
temperature is reached.

The raw materials used in this process must therefore possess a liquid
phase of high viscosity, and the most suitable materials are glass and
plastic. Figure 2 shows the variation of viscosity with temperature for a
typical glass: it will be seen that a definite melting point cannot be
attributed to the material but that the viscosity decreases uniformly
with temperature over a wide range. There are therefore four practical
configurations for an optical fibre:

1. Glass core plus glass sheath.
2. Glass core plus plastic sheath.
3. Plastic core plus plastic sheath.
4. Plastic core plus glass sheath.

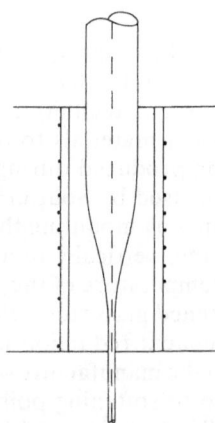

Fig. 1   Illustration of the drawing down of stock material into a fibre within the muffle
furnace.

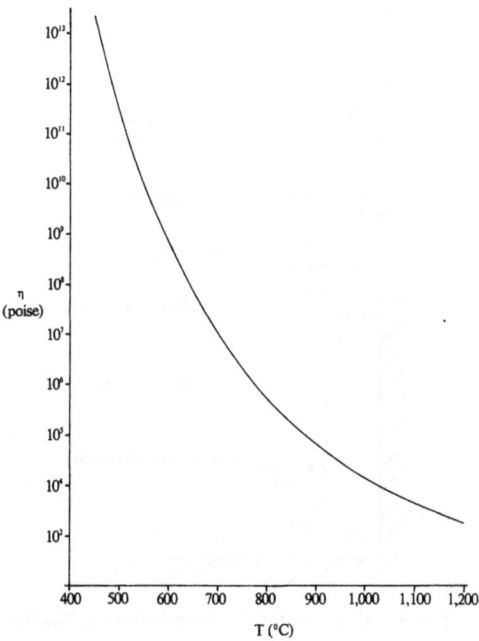

Fig. 2 Variation of viscosity (η) with temperature for a typical fibre optics core glass – Schott F2. (Courtesy of Schott and Gen.).

Of these, the first and third are the only two generally used. In these, the core and sheath will have similar softening points which simplifies the production process.

The basic drawing technique was developed in the early 1950s by Hopkins[3] and van Heel[4] to produce fibres of optical glass. However, these fibres did not have a glass sheath and gave an unsatisfactory optical performance. The fibres drawn by Hopkins were in fact unsheathed, using air as the lower-index medium, and those of van Heel were sheathed with plastic applied by dip-coating a glass fibre in a solution of the plastic material. In 1958, Hirschowitz[5] reported the development of a technique for the production of optical fibres in which both the core and the sheath were glass. Production of optical fibres at the present time uses a development of this technique in the majority of instances.

A. Production of Single Optical Fibres

The production technique for the manufacture of glass optical fibres is shown schematically in Fig. 3. In this, a rod of the core glass is sur-

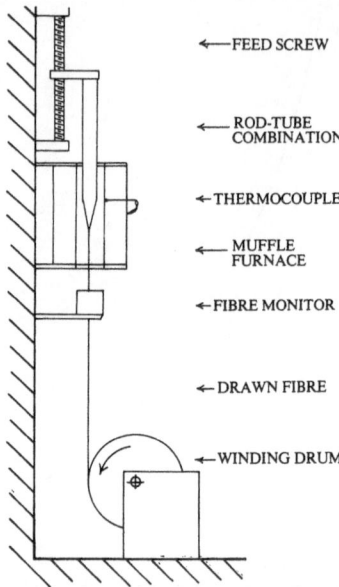

FEED SCREW

ROD-TUBE COMBINATION

THERMOCOUPLE

MUFFLE FURNACE

FIBRE MONITOR

DRAWN FIBRE

WINDING DRUM

Fig. 3   Layout of production unit for the manufacture of flexible optical fibres.

rounded by a tube of the sheath glass, and the combination is mounted in a vertical muffle furnace. The construction of the furnace is such that in operation it exhibits a temperature gradient along its length, giving a maximum temperature in a region near the centre of the muffle, which is monitored by a thermocouple whose output operates the temperature-control circuit. This maximum temperature is defined as the working temperature of the furnace.

To start the process, the rod–tube combination is set up so that its lower end is just inside the maximum temperature zone of the furnace, and the furnace power is switched on. When the working temperature, typically 1000°C, is reached, the glasses soften and a fibre can be drawn from this softened end. Conditions within this region of the furnace must be such that the sheath tube collapses and fuses on to the rod cleanly, to give a good interface. With some glass combinations it is usual to establish a vacuum between the core and the sheath glasses to create a differential pressure across the sheath glass which assists the fusion. This vacuum should not be too high, since this would cause out-gassing of the glasses at the drawing temperature, with the risk of forming gas bubbles on the interface which would interfere with the reflection process. A pressure of around 1.0 mm Hg normally proves sufficient to give good fusion without out-gassing problems. To maintain the cross-section of the fibre constant, the rod–tube combination is fed into the furnace at a constant rate to replace the material being removed from the lower end, and the fibre is drawn off at a constant speed. If these speeds are respectively S and s cm/s it is a simple matter to prove that, if D and d are the diameter of the starting billet and the fibre respectively, these speeds are related by the following equation:

$$s = S \times \frac{D^2}{d^2} \tag{1}$$

This follows from the fact that at equilibrum and in unit time the volume of glass being fed into the furnace must equal that being removed.

Rods of fairly large cross-section can be fed into the furnace at relatively high speeds, since the temperature gradient in the furnace causes a gradual heating of the rod as it moves towards the drawing zone. Typically the rod diameter would be 30 mm and a feed rate of 1 cm/min could be used, and under these conditions the drawing speed required to produce 20 μm fibres would be around 2000 m/min. In order to obtain such high drawing speeds, the fibre is attached to a drum which is rotated at a constant speed, the drawn fibre being stored on the drum until the run is completed. This creates a slight problem since the stored fibre increases the effective diameter of the drum, so that with a constant rotational speed the actual drawing speed increases as the run progresses. However, the increase in effective diameter during a typical production run is only about 1% and this means a diameter change in the fibre of only 0.5% from Eq. (1), which would not be serious. Obviously, the drum diameter must be such that the fibre can be wound without risk of breakage and, since drum diameters greater than a couple of metres are impractical, the maximum fibre size which can be drum wound is around 150 μm. Above this size the fibre is drawn by being passed through a set of friction rollers, or caterpillar tracks, which are driven at a constant speed. As the drawing speed is dependent on the square of the fibre size the values of drawing speed (below approximately 30 m/min) required for such fibres are significantly less than the speed mentioned above, so that no problems normally arise when drawing fibres in this manner. In either method the furnace temperature must be closely controlled since a variation in the working temperature causes a variation in the position of the start of the drawing zone, and this alters the effective value of the feed rate, which is the speed of the glass relative to this zone. Similarly, the furnace should not be subject to draughts which would give rise to a similar effect.

Normally the fibre size is monitored during the run by standard gauging techniques, and a non-contact method is to be preferred to avoid possible damage to the fibre surface. A suitable monitor is illustrated in Fig. 4; this has been used to monitor fibre sizes down to $10\,\mu$m. In this monitor the image of a slit is formed at the plane of the fibre and by reflecting the light from a rotating tilted mirror the image is made to oscillate with an amplitude significantly greater than the fibre diameter. The light is then collected by a lens and focussed on to a detector which is placed at the focal point of the lens, to minimize the effect of variations in sensitivity across the active area of the detector. As the image passes over the fibre some of the light is prevented from reaching the detector, the amount being determined by the diameter of the fibre. The drop in signal level from the detector is a measure of the fibre diameter and can be displayed on a meter. The output from the measuring head can be used in a servo system which corrects variations in diameter by altering the drawing rate. Although theoretically the feed rate could be controlled, this method would have a very long time constant and proves to be impractical.

B. Choice of Raw Materials

The process described above can produce optical fibres of high quality in a range of diameters from 7 μm to 10 mm. In order to obtain the highest quality of optical fibre the raw materials used must be chosen with care. Although the range of refractive indices possible using available glasses is large, and could give a range of N.A.s varying almost continuously from zero to greater than 1.4, the range from which satisfactory fibres can be drawn is much smaller than this. The reason is that the choice of a pair of glasses is not controlled solely by the desired values of refractive index, since there are additional conditions which the glasses must satisfy before strong fibres of good quality can be produced. The main conditions which must be met are as follows:

   1. The softening temperatures of the glasses should be similar. This is essential since the glasses are both heated to softening

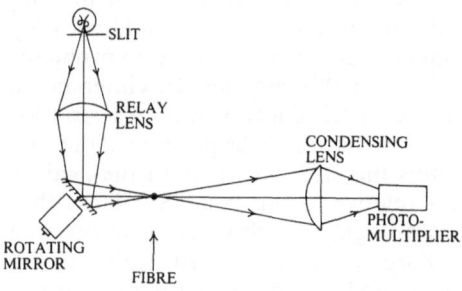

Fig. 4   Non-contact monitor for fibre diameter.

point in the same furnace. If the softening points differ greatly the correct drawing conditions become very critical.

2. Ideally the thermal expansion of the core should be greater than that of the sheath. This ensures a surface compressive strain when the fibre has cooled which greatly increases the strength of the fibre.

3. The glasses must be chemically compatible so that when they come into contact at the drawing temperature there will be no adverse reaction at the interface. This adverse reaction could take one of two forms. First, the glasses may mix slightly at the interface and create an unstable product which will devitrify. This devitrification would create a layer of crystallites at the interface which would seriously degrade the reflection process. Second, this glass mixture may separate into two immiscible phases, in which one phase exists as globules immersed in a matrix of the other phase. Since these phases will inevitably have different refractive indices, a scattering layer will be present at the interface, which will reduce the overall transmission of the fibre.

The first two conditions can be met fairly readily by reference to the properties of the glasses published by the manufacturers. However, the third condition is not so easily satisfied, since little is known of the effects on glasses of contact with other glasses at the temperatures used in the manufacture of optical fibres. At present, glasses are chosen by experiment since the result cannot always be predicted by theory. In a number of cases any adverse reaction is visible under a microscope and the elimination of these combinations is straightforward. However, in other cases the effect is sub-microscopic and shows itself as a slight reduction in transmission or strength. Figure 5 shows an example of phase separation in which the size of the scattering centres is 0.13μm.

When all these conditions are met, it is found that only a few glass combinations can in fact be used. Fortunately, these yield a fairly large range of N.A.s from zero up to 1.2 approximately. This is not a continuous range however; the majority of combinations give N.A.s between 0.4 and 0.7. A further practical restriction exists since optical glasses are not available in the form of tubing and most manufacturers tend to use commercial glass tubing as the sheating glass. The range of softening points and thermal expansions available in this type of glass is limited, which imposes further limitations on the choice of a matching core glass.

Since the optical quality of the interface must be very high, the quality of the initial rod surface and the inner surface of the tubing must be high. These surfaces are thoroughly cleaned before drawing,

Fibre Optics

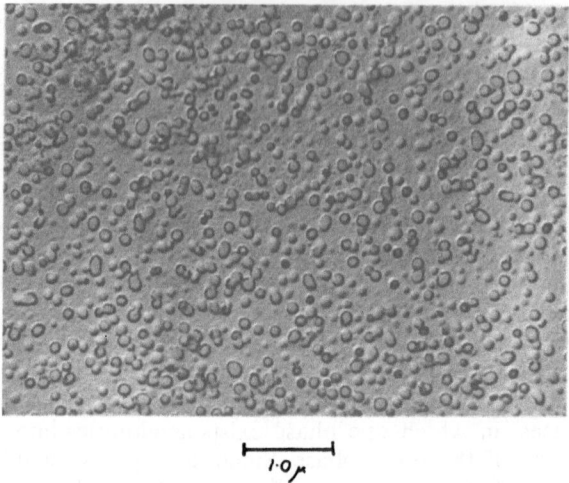

Fig. 5 Electron micrograph showing phase separation in heat-treated glass (×
30,000). (Courtesy of Sheffield University, Dept. of Glass Technology).

and should not have deep pits or scratches. Minor imperfections tend
to disappear through the action of the drawing process and the "fire
polishing" which takes place in the drawing zone. In the present state
of the art the losses incurred owing to an imperfect interface are
approximately equal to those due to the absorption of the bulk glass.
If a significant increase in the overall transmission of optical fibres is
to be gained, then both sources of loss must be considerably reduced.
Since glasses are now being developed specifically for optical fibres,
work has been done in recent years on the reduction of interfacial
losses.

One promising technique involves the extrusion of the fibre from
molten glass using two concentric crucibles, the inner one containing
the core glass and the outer one the sheath glass. The interface is
formed in the extrusion dies just before the fibre is drawn and should
therefore be free from contamination. The problems which can arise
from the mixing of the two glasses will be aggravated since the temperatures involved are much higher than in the normal method of
manufacture, but these can be avoided through the careful choice of
glasses. This will be discussed in more detail in Chapter 10 when dis-
cussing fibre optics in the infra-red region of the spectrum. Another
approach involves the production of a fibre which has no sharp interface. Instead, the refractive index of the system decreases uniformly
with increasing distance from the centre so that the guidance is

achieved by refraction rather than reflection. The theory of operation of such a system is outlined in Chapter 10.

II. OPTICAL PROPERTIES

The theoretical considerations of the propagation of light through an optical fibre did not include practical losses, for reasons of simplicity. However, in any practical system these losses must be considered. They can be itemized as follows:

A. Fresnel Reflection Losses

These losses are incurred at the input and output faces of the fibre and are due to the difference in refractive index between the core glass and the immersion medium. Figure 6 shows the reflection losses versus angle of incidence for a glass with a refractive index of 1.62, for a single air–glass interface and for the two directions of polarization. It will be seen that for a large range of angles the equation for normal incidence

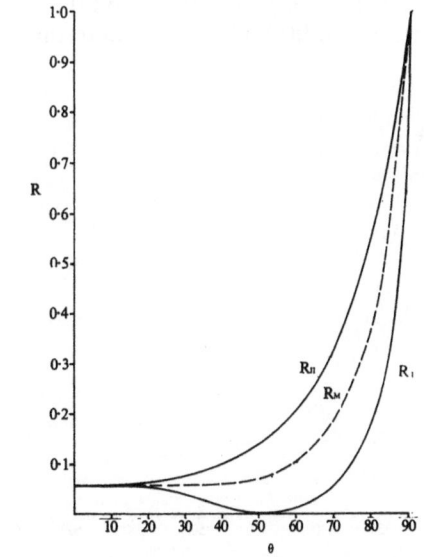

Fig. 6   Variation of the coefficient for Fresnel reflection with incidence angle (θ) for (a) light polarised perpendicular to the plane of incidence ($R_\perp$), (b) light polarized parallel to the plane of incidence ($R_{//}$), (c) unpolarized light (R_M)($n = 1.62$).

is sufficiently accurate, i.e.:

$$A_F = \frac{(n_1 - n_0)^2}{(n_1 + n_0)^2} \qquad (2)$$

where n_1 and n_0 are the refractive indices of the core and immersion medium respectively, and A_F is the fraction of the incident light which is reflected.

B. Absorption Losses

The core glass is not perfectly transparent and the transmission through a length L of the core glass is given by:

$$T_\alpha = e^{-\alpha L} \qquad (3)$$

where α is the absorption coefficient. The path length, L, will be calculated from Eq. (10) of Chapter 2. Figure 7 shows the variation in α with wavelength for a typical core glass, Schott F2.

C. Interface Losses

If there are imperfections in the interface, or if the sheath glass is absorbing, then the reflection process will not have an efficiency of unity. Potter[6] has measured the reflection coefficient for a number of optical fibres and finds a value of around 0.9995 for glass fibres, and 0.99 for plastic. An additional factor in determining the efficiency

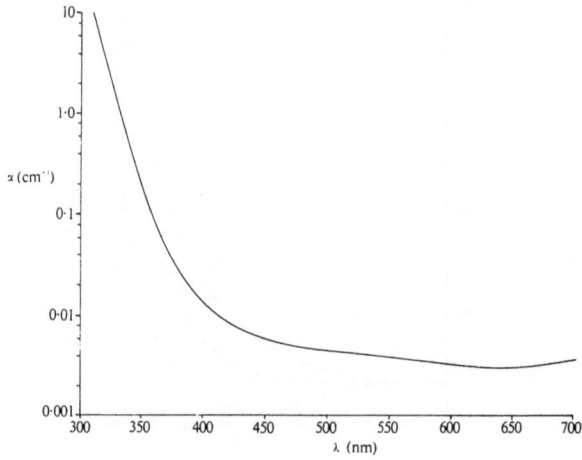

Fig. 7   Variation of absorption coefficient (α) with wavelength for a typical core glass – Schott F2.

of the reflection process is the thickness of the sheath material. In some systems the fibres are embedded in an absorbing material and this will reduce the reflection efficiency to an extent determined by the thickness of the sheath material, since the energy distribution in the light beam is not confined within the core (section 2, IV).

D. Fibre Transmission

The number of reflections suffered by a ray, it will be recalled from Eq. (12) of Chapter 2, is dependent on the angle of incidence. For any given angle, the transmission of a fibre can be represented as:

$$T_f = (1 - A_F)^2 e^{-\alpha L \sec\theta'} (1 - \beta)^{\eta L} \tag{4}$$

where A_F and α are defined above, $L \sec\theta'$ is the path length in the fibre, β is the reflection loss and η is the number of reflections/unit length. Since $\beta \ll 1.0$ we can write (4) as:

$$T_f = (1 - A_F)^2 e^{-\frac{L}{\cos\theta'} \left(\alpha + \beta \frac{\sin\theta'}{d}\right)} \tag{5}$$

where we have substituted from Eq. (11) of Chapter 2 for η. Thus the transmission of an optical fibre will vary exponentially with length for any given input angle, with an apparent coefficient of absorption given by:

$$\left(\alpha + \beta \frac{\sin\theta'}{d}\right) / \cos\theta'$$

The loss per reflection can be regarded as being made up of two parts. The first part consists of the energy lost by scattering at an imperfect interface and the second consists of the absorption or scattering of the energy due to its finite penetration into the sheath. This latter is discussed in detail in Chapter 9. Figure 8 shows the variation in transmission with length for a typical optical fibre at a wavelength of 500 nm.

The absorption coefficient of the core glass varies with the wavelength of light used, and it is this which determines the wavelength range over which useful transmission may be achieved. Figure 9 shows the spectral transmission of an optical fibre 50 cm and 100 cm long. It can be seen that a useful transmission may be expected from about 400 nm to 1200 nm. This range is typical of a fibre made using an optical glass core, whilst with a plastic fibre the transmission curves are as shown in Fig. 10. In these measurements, a collimated beam of light is passed through the fibre and the amount transmitted is measured by a suitable detector. This avoids any complication due to the beam exceeding the critical angle for reflection owing to bends in the fibre, variations in diameter etc. Because of this tendency towards de-

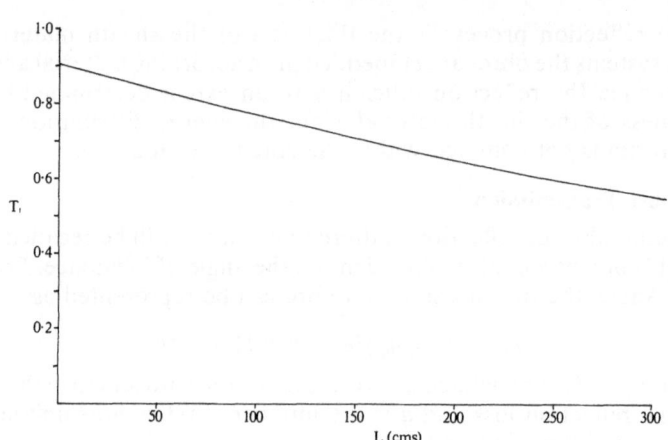

Fig. 8   Variation of transmission (T_f) with length (L) for a typical optical fibre.

collimation, the optical system which condenses the light on to the detector should have a higher N.A. than the input optics, and this should ideally equal that of the fibre. It can be seen then that the normal optical fibre functions efficiently in the visible region of the spectrum. Special optical fibres have been made for use in the ultra-violet and infra-red regions of the spectrum; these will be discussed in a later chapter.

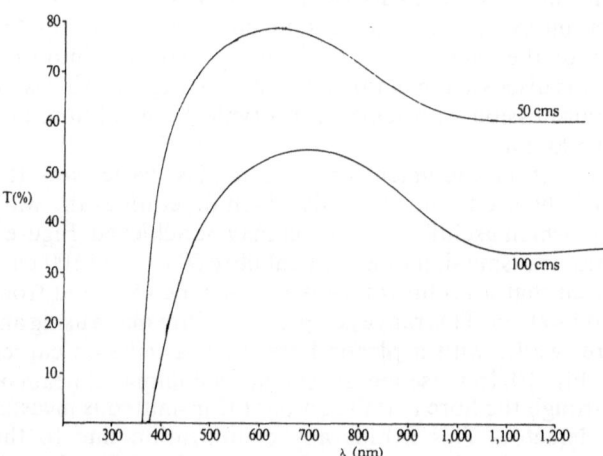

Fig. 9   Spectral transmission of a glass optical fibre 50 cm and 100 cm in length.

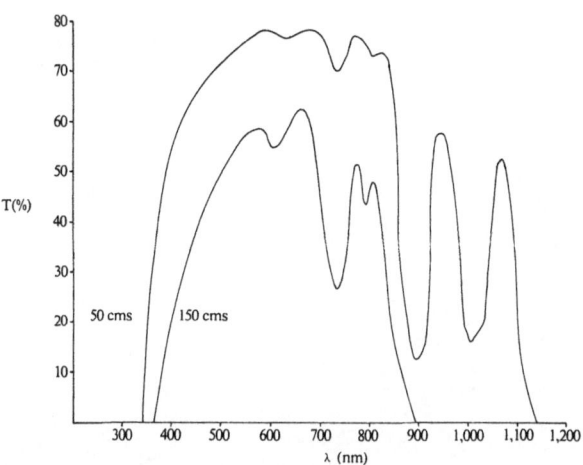

Fig. 10   Spectral transmission of a plastic optical fibre 50 cm and 150 cm in length.

III. STRENGTH OF AN OPTICAL FIBRE

The main physical property of an optical fibre which is of interest is its mechanical strength. This determines to a large extent the degree of bending to which a bundle of such fibres may be subjected, and thus influences the flexibility and ease of handling of the whole system.

Glass exhibits a strength which is several orders of magnitude less than that predicted theoretically. The reason for this marked reduction in strength is now generally believed to be the presence of minute flaws on the glass surface which act as stress multipliers, causing local stresses some 10 to 100 times the overall applied stress. When these local stresses reach the theoretical breaking stress, rupture of the glass commences. The theory of these was first put forward by Griffith[7] in 1920 and the flaws are called Griffith flaws. These flaws cannot be seen directly but recently a technique has been developed by Ernsberger[8] which renders them visible. Basically, the glass surface is subjected to tensile stress acting parallel to the surface which opens the flaws to give visible surface cracks. This is achieved by a base-exchange process in which sodium ions in the glass are replaced by smaller lithium ions. An etching process is then used to render the resulting cracks more visible; Fig. 11 shows the results of this treatment when applied to a typical sample of glass. The increase in the density of the Griffith flaws in the contaminated region is obvious.

The Griffith flaw theory accounts for all the known facts concerning

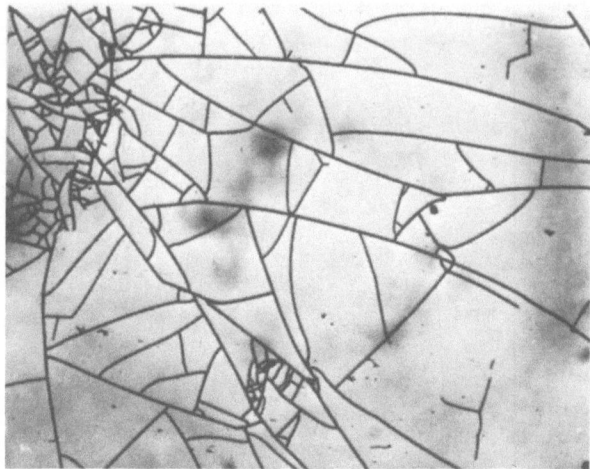

Fig. 11   Photomicrograph (× 90) of Griffith flaws in a glass surface, made visible by the Ernsberger technique. The region in the top left hand corner has been impacted by a hard particle. (Courtesy of Dept. of Glass Technology, Sheffield University)

the strength of glass. Those which are of relevance to optical fibres will now be discussed.

Since the rupture of glass is primarily a surface effect, one would expect that the strength of a fibre will increase as the diameter is decreased. This is found to be so, and Figure 12 shows the variation in strength of glass fibres with their diameter. The reasoning behind this is that, for a given length of fibre, the surface area, and therefore the total number of Griffith flaws, is directly proportional to the diameter. By extrapolation of the curve in Fig. 12 to zero diameter, a figure for the strength with zero surface area, i.e. no flaws, is found (1.1×10^5 kg/cm^2) which agrees with the theoretical strength. This curve was constructed from values obtained by Griffith. Similar reasoning to the above indicates that the strength of a fibre will decrease as the length of sample increases.

The actual density of these flaws on the glass surface is not a characteristic of the material itself, but is determined by the preparation and treatment of the surface. A freshly-prepared surface has very few flaws, but abrasion or contamination of the surface creates large numbers of flaws, (see Fig. 11). A freshly prepared fibre is therefore very strong, but its strength decreases as a function of time and amount of abrasion. The effect of the latter can be significantly reduced by lubricating the surface of the fibre. The decrease in strength with time seems to be attributable to the presence of moisture in the air sur-

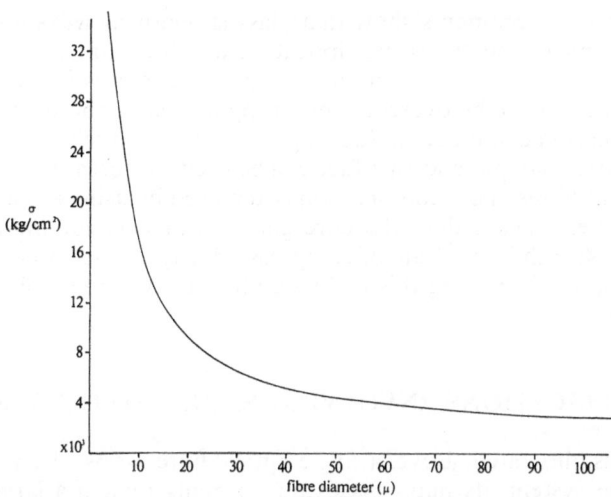

Fig. 12   Variation of the tensile strength (σ) of glass fibres with diameter.

rounding the fibre which reduces the surface energy. The strength of fibres measured *in vacuo* is about three times that obtained in moist air. The values for breaking stress shown in Fig. 12 indicate that glass fibres are some ten times stronger than bulk glass, for the reasons already discussed. Because of the statistical nature of the mechanism of glass failure a wide variation in strengths can be seen in measurements on similar samples. Figure 13 shows the distribution of strengths obtained with similar samples. It will be seen that variations in strength of up to a factor of two can be expected from typical sample batches.

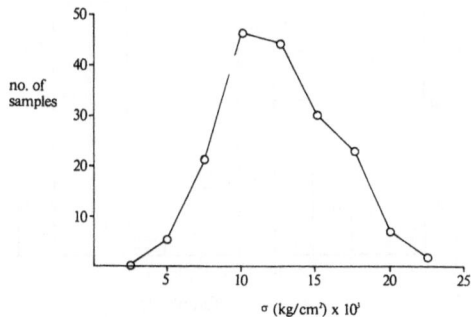

Fig. 13   Distribution in tensile strengths obtained from tests using fibres with the same diameter and length.

The above arguments show that glass is relatively weak under tension, owing to the stress multiplication which occurs at the Griffith flaws. If the surface could be placed under a compressive strain then this would have to be overcome by an applied tension before this stress multiplication could occur. Such a glass would be much stronger than a stress-free sample and this fact can be used to increase the strength of optical fibres. This compression is achieved by using a sheath glass of lower expansion than the core glass, so that on cooling the core pulls the sheath into a state of compression. Figure 14 shows the effect on strength of applying this technique to a conventional fibre-optics core glass.

IV. APPLICATIONS INVOLVING SINGLE OPTICAL FIBRES

As was illustrated above, a single optical fibre forms a very efficient guidance system, its main disadvantage being that, if a large cross-section is required, the system cannot be flexible. However, there are several applications for single fibre guides in which the unique properties of optical fibres are used to advantage.

Single fibre guides are made by suitably finishing the ends of a single optical fibre and in the majority of applications this means grinding and polishing the end-faces of the fibre normal to the fibre axis. If the fibre is relatively large in diameter, greater than 1.0 mm say, then conventional optical working techniques can be used. If the fibre diameter is less than this it must be supported during processing, and the major fibre-optics companies have developed techniques for grinding

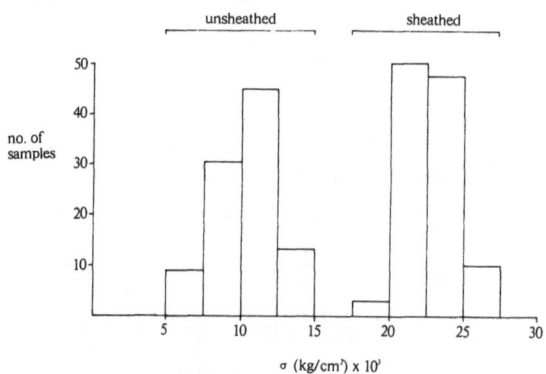

Fig. 14   Illustration of strength increase achievable by the use of a suitable sheath material.

and polishing fibres already mounted in the finished component. This requires special processes since these fibres will be held in position using an epoxy resin, and the finished surface may be flush with a metal or plastic support member.

A. Sampling

A flexible single fibre makes an extremely useful sampling device for selecting and measuring the light intensity of a small region within an illuminated area. An example of an application of this nature is illustrated in Figure 15, which shows diagrammatically the assembly of an image probe used in the evaluation of lens systems. The input face of an optical fibre is placed in the system under test and can be made to traverse at right angles to the axis of the system in the image plane of a bar test chart. The light issuing from the fibre is fed into a photo-detector system, and the output from this is recorded during one traverse. Information regarding the optical performance of the lens system can be deduced from this record. The fibre diameter can be as small as 10μm so that reliable information can be gained with test-chart frequencies up to 20 lp/mm.

This application utilizes the flexibility and high transmission of the optical fibre. A further advantage is that the output light is uniformly distributed over the fibre face, owing to the effect of skew rays. This eliminates spurious effects due to non-uniformity in response across the photo-detector face, which can occur when the fibre input is traversing the boundary between the bright and dark regions of the image. If necessary, the fibre input can be attached to a vibrating member to give high-speed sampling, although in this case the motion of the end would be sinusoidal and it would be necessary to use only the centre portion of the sweep if a linear scan were required.

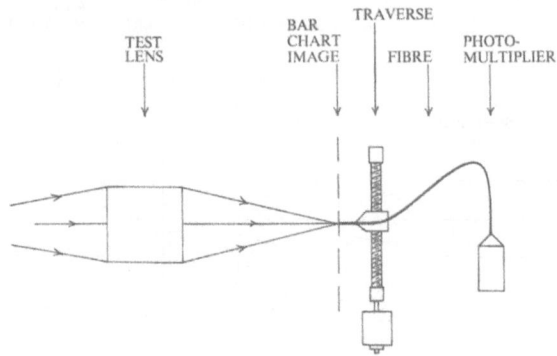

Fig. 15   Diagram of the image probe.

A novel extension of this idea has been used to manufacture an alignment device which was intended to detect when a remote light source was on the axis of the device. This is shown diagrammatically in Fig. 16. In this example the fibre itself is used as the vibrating member since the resolution required allowed the use of a fairly rigid fibre (about 250 μm in diameter). The vibration was caused by mounting a small piece of soft iron at the fibre tip and placing this between the poles of a magnet giving an asymmetric field, driven by an alternating current. The fibre was clamped at a point such that the resonant vibrational frequency of the fibre – iron system equalled the frequency of the electrical supply, and vibrational amplitudes of 5 mm to 10 mm could be achieved. If a higher resolution is required, then the core diameter can be reduced while maintaining the overall fibre diameter constant.

B. Small Apertures

One recurring problem in a number of areas within optics is that of providing a small aperture of high transmission. The main reason for low transmission is the difficulty of making small holes satisfactorily in thin material. In addition, the material thickness adjacent to the hole must be small compared to the aperture diameter, so that a reasonable solid angle of light can be transmitted. Obviously, a single fibre fulfils this task admirably, since a high beam angle can be maintained, independent of physical dimensions. Applications for this range from multiple apertures in a Nipkov disc to a small light source for the optical evaluation of mirror performance.

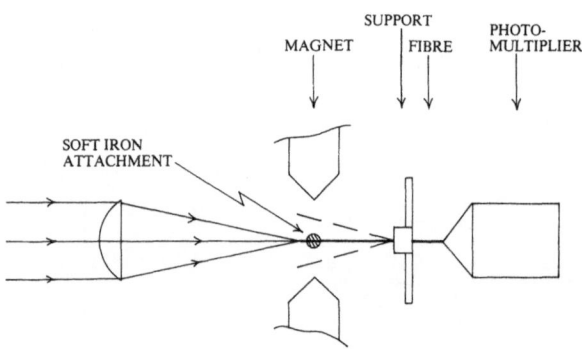

Fig. 16   An alignment instrument using a vibrating fibre.

C. Channelling of Heat

Single fibre guides have been used successfully to transfer the infra-red energy from the image of a source to the point of application of the heat. Although some of the radiation is absorbed by the glass, useful amounts of heat can be transferred over distances of about 30 cm from medium-power lamps, e.g. a 100-watt Tungsten Halogen lamp.

In a practical example of this type, the heat from such a lamp was guided via a single optical fibre, 2 mm in diameter, over a distance of 25 cm. This was used to make soldered connections to the base of a pin in a multiple connector. The fibre system had two advantages over the other systems tried, a soldering iron and a conventional light-condensing system. First, the heat was localized so that connections already made on neighbouring pins were not affected. Secondly, the system did not require focussing and was compact; it could be used in a similar fashion to the conventional soldering iron. This system was not flexible, which would have been desirable, but a flexible fibre bundle could not be used since the temperature at the image plane of the source was in excess of 1000°C, which would have completely destroyed the resin used to consolidate the fibre ends. However, owing to the low absorption of the single fibre, the end of the guide did not achieve temperatures much in excess of 150°C, well within the safe upper limit for the glasses used.

D. High-Temperature Environment

Since the single fibre is solid glass, it will not be affected by temperature to the same extent as a flexible fibre bundle. Although the absorption of glass increases with increasing temperatures, useful transmissions can still be obtained in lengths up to 30 cm at temperatures around 500°C where, obviously, the glasses chosen should not soften at the operating temperature. In the majority of such uses,[9] a single fibre is employed to guide the light to a region of lower temperature where it is coupled to a flexible system. Typical applications include the monitoring of the temperatures of jet-engine exhausts, illumination inside ovens, and measurements on hot steel strip.

E. Integrating Chambers

Because of the effect of skew rays and the nonimaging property of the optical fibre, the distribution of light over the output face of the fibre will be uniform if a large number of reflections is experienced by the beam, even if the light distribution across the input face is highly non-uniform. Thus a single optical fibre can be used as an integrating chamber to even out a non-uniform beam of light. There are two main areas of use for an optical fibre operating in this manner. First, it is used when a non-uniform beam is to be fed to a photo-detector where variations in sensitivity over the detector surface may prove troublesome. Secondly, it is used to combine the outputs of a number of systems, either conventional or fibre-optics ones, on to a common optical axis, so that these may pass through a common optical system. Use has been made of this in colorimetry and display applications which will be discussed later.

It can be seen from the above that the single optical fibre is a useful optical system in a number of specialized cases, where relative rigidity is acceptable or where a small-diameter system is essential.

REFERENCES

1. H. S. Soutter, *Proc. Phys. Soc. (London)* **24**, 166 (1912)
2. F. H. Newman and V. H. L. Searle, *The General Properties of Matter*, Edward Arnold, London (1957), p. 167
3. H. H. Hopkins and N. S. Kapany, *Nature* **173**, 39 (1954)
4. A. C. S. van Heel, *Nature* **173**, 39 (1954)
5. B. I. Hirschowitz, L. E. Curtis, C. W. Peters and H. M. Pollard, *Gastroenterology* **35**, 50 (1958)
6. R. J. Potter, *J. Opt. Soc. Am.* **51**, 1079 (1961)
7. A. A. Griffith, *Phil. Trans. Roy. Soc.* **A221**, 163 (1920)
8. F. M. Ernsberger, *Proc. Roy. Soc.* **A257**, 213 (1960)
9. H. M. Runciman, W. B. Allan and I. M. Ballantine, *J. Sci. Instr.* **43**, 812 (1966)

CHAPTER 4

Non-Coherent Bundles – Manufacture and Properties

In the previous chapters we have discussed in detail the physical and optical properties of the optical fibre. However, the major applications of the technology of fibre optics utilize the properties of assemblies of such fibres. The next few chapters will be devoted to a discussion of these properties and the techniques involved in manufacturing such assemblies, or bundles as they are generally called.

These applications can be divided into two main classes. In the first of these, the bundles are used simply as a means of guiding beams of light and in the second, use is made of the optical isolation of an optical fibre to produce bundles in which each fibre acts as an independent channel, so that information about the spatial distribution of intensity in the light beam can be transferred through such bundles. Bundles belonging to this latter class are known as coherent bundles, whilst those belonging to the former are known as non-coherent bundles.

In this chapter we will consider the first class of bundles, which are known simply as light guides.

I. CHARACTERISTIC PROPERTIES OF OPTICAL-FIBRE BUNDLES

In common with any optical system, an optical fibre has the property of being able to transfer light between two points. There is, however, only one set of conjugate planes for the optical fibre system which are fixed and coincident with its end-faces. In the great majority of applications the fibres are in the form of straight cylinders, and under these conditions the angular magnification is unity. This system differs from an equivalent lens system in that it tends to smooth out any variations in the azimuthal distribution of light due to the effect of skew rays, as discussed previously. Similarly, any variations in illumination across the input face are smoothed out and the output from the system is normally a uniform disc of light with a diameter equal to that of the fibre core.

An optical fibre system can posses certain characteristics which distinguish it from a more conventional system; these are:

A. Flexibility

The condition for the acceptance of a ray of light by the optical fibre is independent of the physical dimensions of the fibre. Therefore the fibre diameter can be made sufficiently small to permit flexing; for example, a 50 μm fibre can be bent to a radius of 1.0 mm without breakage and with a negligible effect on its transmitting efficiency.

B. High Beam Angles

The maximum axial angle at which a light ray will be accepted into the system is controlled solely by the refractive indices which define the optical fibre. This angle can be made much larger than that which would be possible in a conventional system, and optical fibres can be made to accept light incident on the end-faces at angles up to 90° to the fibre axis.

C. The Splitting and Combining of Light Beams

A bundle of optical fibres will behave as an assembly of independent guidance systems, so that the spatial relation between the fibres need not be constant. It is possible therefore to have the fibres arranged to provide a number of separate inputs and outputs.

D. Alteration of the Cross-Section of Light Beams

Because of the independence of operation of the optical fibres comprising a bundle these can be assembled with different configurations at the input and output faces, so that the cross-section of the light beam is altered on its passage through the bundle.

The above characteristics follow directly from the properties of the fibre system, which has been discussed in detail in Chapter 2. In addition, there are two other characteristics of this system which are indirect consequences of the theory; these are:

E. Increased Transmission of the Overall System

By utilizing the greater design freedom permitted by the optical-fibre system, it is normally possible to produce a system with a shorter path length or fewer glass-to-air interfaces than the conventional design; both of these modifications yield an increased transmission. In some cases it is also possible, using optical fibres, to increase the solid angle of light transmitted through the system which again yields an effective increase in transmission.

F. Reduction in the Physical Size of the System

The use of flexible optical fibres can lead to a much more compact design owing to the ease with which such a system may be folded. In addition, the dimensions of the system are then in no way influenced by the angular divergence of the light beam. Both of these factors lead to a reduction in the physical size of the system.

Any valid application of fibre optics must utilize at least one of the above characteristics.

II. MANUFACTURE OF LIGHT GUIDES

The term 'light guide' is now generally accepted as defining an assembly of a number of optical fibres mounted and finished in a component which is used solely to transmit light flux.

The high optical efficiency of optical fibres plus their flexibility suggests applications where both these properties are utilized. However, a single flexible fibre has too small an active area for most of these applications and, to retain flexibility with a large active area, a number of such fibres are used in parallel and form a light guide.

The optical fibres are produced as described previously by drawing the fibre on a rotating drum.[1] The run is stopped when the desired number of turns have been completed, and the fibres removed by cutting through the resulting bundle on the drum. Ideally, the drum perimeter should equal the required bundle length plus a small excess to allow for finishing. The ends of this bundle are mounted in some form of jig to compact the fibres and these are bonded together with an epoxy resin; the consolidated regions can then be ground and polished to give plane end-faces. To facilitate the mounting of the finished component the ends may be mounted in a metal or plastic ferrule and, in this case, the ferrule is used as the compacting jig. In addition, the flexible portion of the light guide can be encased in a flexible trunking which protects the fibres. Components for general industrial use incorporate both of the above variations which yield a very robust component, capable of withstanding most working environments (see Fig. 1). In these the trunking is of convoluted metal which limits the bending radius to a value in excess of that which will damage the fibres.

Where a large number of similar light guides have to be made, it is wasteful to have to stop the drawing run at the finish of each bundle, and, in a production process, the winding drum can be indexed along its axis to collect a fresh bundle without stopping the run. This indexing action is triggered by a pre-set counter which counts the drum revolutions, and is set to the required number of fibres at the start of the run. If the fibre diameter varies throughout the run, however, this technique will produce a bundle of the wrong size. In a refinement of this technique, the output from the diameter monitor is fed into an electronic system which squares and integrates the diameters to give a measure

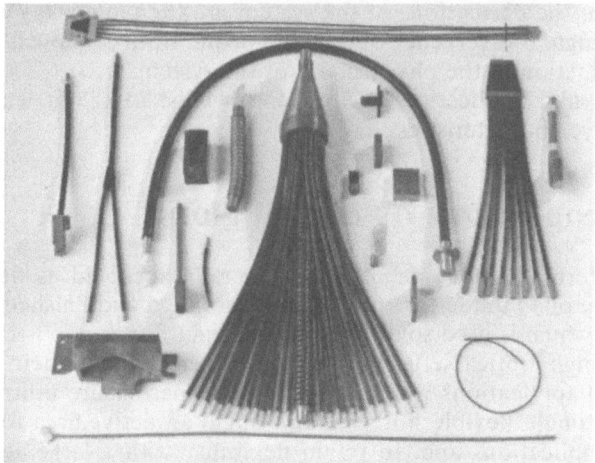

Fig. 1 Photograph showing finished light guides for industrial use. (Courtesy of Rank Precision Industries Ltd)

of the cross-sectional area of bundle already wound on the drum, and the indexing is triggered when this reaches a pre-set value.

Although many applications use light guides which have the same cross-section at each end, this is not mandatory since the end-face is built up from a large number of fibres which need have no specific spatial relation with each other. Indeed, many applications exist in which the cross-sections differ and almost any shape can be made, the only criterion being that, since there are the same number of fibres at each end, the end-face areas must be equal. Similarly, there is no need to have a single input or output and branched light guides can be made where the only criterion to be met is that the sum of the input active areas must equal that of the output active areas. This criterion is only approximately true since it assumes that the fibres are infinitely small; however it is sufficiently accurate for all known applications. Obviously, by utilizing these variations, a very large number of components can be designed which are all described under the generic heading of light guides. The manufacturing processes for these are basically the same as that already described for the case of a simple light guide (i.e., one input and one output face) with circular ends.

These more complicated components can perform functions other than a simple transmission of energy since they can change the shape of a beam of light or split it into a number of beams. In some applications, these functions are the prime ones and flexibility may not be desired, so that the whole component can be potted in some rigid enclosure, e.g. a plastic moulding, which makes the final component extremely rugged.

A. Multiple Production of Optical Fibres

The process just described is the conventional method of producing light guides. However, it has certain disadvantages as a production process; these are:

1. The maximum length of light guide is governed by the circumference of the drum.
2. If a bundle of fibres is required, then the drum must be rotated through a number of revolutions equal to the required number of fibres in the bundle.
3. Since the length of bundle produced is limited, further processing must be done on a batch basis.

These last two disadvantages mean that the price of the finished component is high and unattractive to customers in mass-production industries. In order to overcome these disadvantages, a different approach to the production of optical fibres was developed.

In this technique the fibres required to make up a bundle are drawn simultaneously from a number of billets, and this bundle is collected, as in the previous method, on a drum from which any length can be unwound at the end of the drawing run. To maintain the bundle as a separate entity on the drum, a fine thread is wound helically around the bundle as it is drawn, with a pitch of about 20 cm. The equipment required to draw such bundles is shown diagrammatically in Fig. 2.

Since the number of billets used in this process equals the number of fibres in the required bundle, which would typically be about 400, the billets have to be much smaller than those used in the production of a single fibre, a typical diameter being 6 mm. These billets are suspended vertically above the furnace system from a frame which is capable of being lowered at a constant rate. It is obviously impractical to have a muffle furnace for each billet, and so billets are grouped together in assemblies of 50 to 100, each assembly being fed into a furnace of a design similar to that described in the previous chapter. Care must be taken to ensure that the softened ends of these billets do not touch and fuse during the drawing process and the simplest way to achieving this is to have a reasonable distance separating the billets in each assembly. This means that in a typical run the billets are spaced over an area of about 1 m square.

The drawn fibres are converged using a series of metal funnels and finally compacted by passing through a felt-lined die. The bundle is then passed through a serving mechanism which winds a thread around

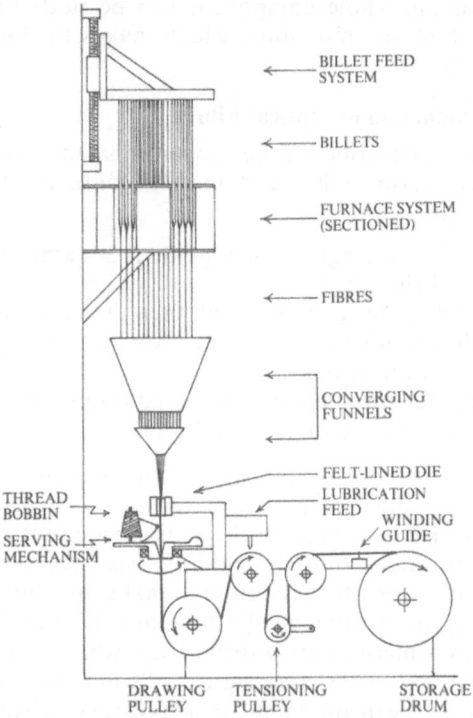

BILLET FEED
SYSTEM

BILLETS

FURNACE SYSTEM
(SECTIONED)

FIBRES

CONVERGING
FUNNELS

FELT-LINED DIE

THREAD   LUBRICATION
BOBBIN   FEED   WINDING
                                  GUIDE
SERVING
MECHANISM

DRAWING   TENSIONING   STORAGE
PULLEY   PULLEY   DRUM

Fig. 2   Schematic layout of machine for the quasi-continuous production of optical
fibre bundles.

the bundle to prevent the fibres in the bundle becoming entangled with
the fibres already wound on the drum. The served bundle passes over
a series of pulleys whose purpose is to control the tension in the bundle
and which incorporates a means for applying lubrication, essential
in this process. The lubricant used is ideally a member of the stearate
family and is applied diluted in a volatile solvent. The bundle is finally
stored on a rotating drum until the run is complete. It should be noted
that the drawing is done by the first pulley in the train and not by the
storage drum, hence the need for tension control. The entire production process is much simplified if the starting billets are prepared so
that the sheath is already fused to the core. This is achieved by producing 6 mm single fibres by the method described in the previous chapter.

Optical fibres can be produced in this manner at much lower cost
than before with no restriction in length; a typical run would produce
around 20 000 m of bundle. In addition, the fact that one can have a
virtually continuous length of bundle opens up the possibility of using

mass production techniques in subsequent processing so that the cost of the finished component is very low. Optical fibres produced in this manner exhibit the same optical and physical properties as those produced by the more conventional technique.

III. PROPERTIES OF LIGHT GUIDES

A. Optical Properties

The transmission curves for a typical range of light guides are shown in Figs. 3 and 4. Figure 3 shows the variation of the transmission for white light with length of light guide, and Fig. 4 shows the spectral variation of transmission for a light guide 100 cm long. As would be expected, these follow the same general form as the equivalent graphs for single optical fibres, but the corresponding values for transmission are lower. This is because the end-faces of a light guide consist of the polished ends of optical fibres which are closely packed and bonded together with some adhesive. When a beam of light is incident on the end-face it is only the light which impinges on the fibre cores which is transmitted through the guide, so that a portion of the light is lost at the input. This is shown clearly in Fig. 5 where the end-face of a light guide is shown in diagrammatic form; the shaded areas are the fibre cores, i.e. the active areas of the end-face.

For hexagonally-packed fibres, as shown in the diagram, it can be

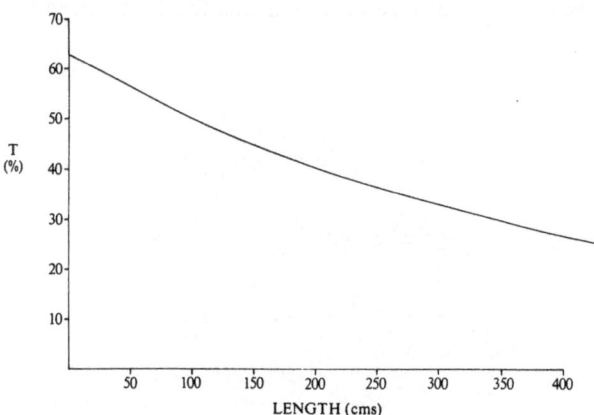

Fig. 3   The variation in the 'white light' transmission (T) of a simple light guide with guide length.

Fibre Optics

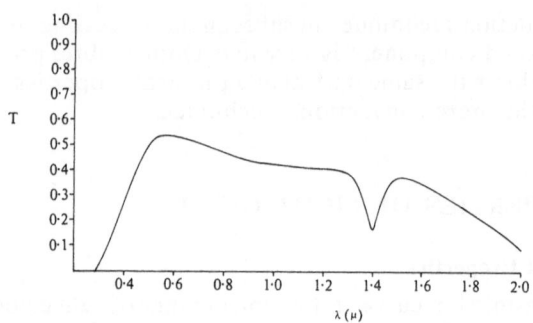

Fig. 4   The spectral transmission of a simple light guide 100 cm long.

shown that the fractional area occupied by cores is given by:

$$F_c = \frac{\pi}{2\sqrt{3}} \frac{d^2}{D^2}$$

$$\approx 0.91 \frac{d^2}{D^2} \tag{1}$$

where D is the overall fibre diameter, and d is the core diameter. The ratio d/D is determined by the basic design of the fibre, and the active area is thus a constant fraction of the total area for a given design, and is independent of actual fibre diameter, provided that this is much less than the end-face dimension. A typical fibre would have a d/D ratio of 0.88 which gives a value for F_c of 0.70. Thus, only around 70% of the incident light strikes a fibre core and this fractional area is known as

Fig. 5   Representation of part of the input face of an optical fibre bundle with the active area, (i.e., fibre cores) shaded.

the core packing fraction. For any light guide the transmission will be given by the transmission of the optical fibre multiplied by this core packing fraction. By extrapolating the curve in Fig. 3 to zero length a transmission of 63% is obtained which is the fraction of light usefully accepted and is equal to the core packing fraction less Fresnel reflection losses.

Light guides are used mainly in conjunction with other optical systems which means that the beam angles are normally below 30° semi-angle. Within this range of angles, it can be shown from Eq. 5 of Chapter 3 that the transmission can be represented as an exponential with constant coefficients, i.e. we can write the transmission as:

$$T_f = A_T e^{-\beta x} \tag{2}$$

where A_T represents end-effects, β is an absorption coefficient, and x is the fibre length.

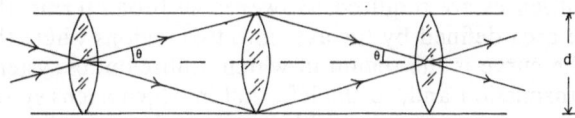

Fig. 6   Lens relay system with diameter, d, capable of transmitting a light beam of semi-angle θ.

It is instructive to compare the transmission efficiency of a light guide, as represented by Eq. (2), with that of a lens relay system, as depicted in Fig. 6. For the sake of simplicity it will be assumed that the absorption of the lens is zero, and that the imagery is perfect. Each lens in the relay will have a transmission T_R which is less than unity due to Fresnel reflections, and the number of lenses (n_L) in a length of x cm for a beam semi-angle θ will be given by:

$$n_L = \frac{2 \tan \theta}{d} \cdot x \tag{3}$$

Thus the transmission (y) is given by:

$$y = (T_R)^{n_L}$$

$$= (T_R)^{\frac{2x \tan \theta}{d}} \tag{4}$$

The transmission of this system will equal that of a light guide when:

$$A_T e^{-\beta x} = (T_R)^{\frac{2x \tan \theta}{d}}$$

i.e.

$$\ln A_T - \beta x = \frac{2x \tan \theta}{d} \ln T_R \tag{5}$$

This relation is plotted in Fig. 7 as the variation of length (x) against semi-angle (θ) for $d = 0.1$ cm and 1.0 cm, assuming that:

$$A_T = 0.63$$

$$\beta = 2.2 \times 10^{-3} \text{ cm}^{-1}$$

and $$T_R = 0.92$$

The value for T_R is that for an unbloomed lens, and Fig. 8 shows the same relation plotted for a bloomed lens, where a value for T_R of 0.98 has been assumed. It is also assumed that the transmission of the lens system varies smoothly with the number of lenses, i.e. that $n_L \gg 1.0$, which means that inaccuracies occur in cases where only a small number of lenses are required. However, in broad terms, the curves divide the area defined by the axes into two regions where that to the right of the curve is the region in which a fibre optics system has the higher transmission and, to the left, that in which a lens system yields the higher transmission. As would be expected, for small beam angles the lens system is the favoured one; however the "fibre optics" region is much larger, except in the case of bloomed lenses and a 1.0 cm channel diameter. The assumptions made regarding the lens system will mean that, in practice, these will perform less efficiently than the above discussion implies.

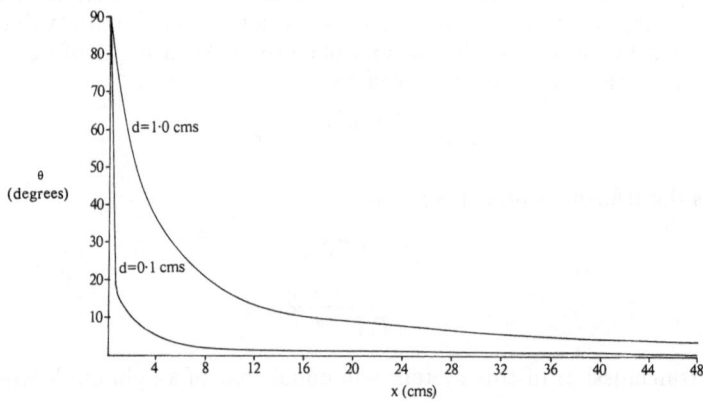

Fig. 7 Diagram showing the loci of points for equal transmission of a simple light guide and an idealized lens system with unbloomed lenses, for two diameters of channel.

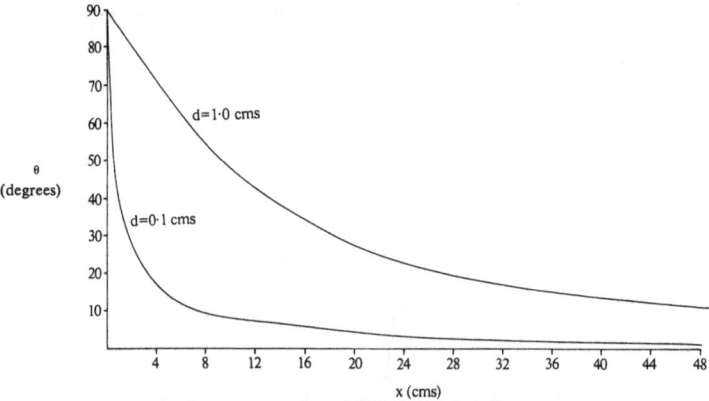

Fig. 8   Diagram showing the loci of points for equal transmission of a simple light guide and an idealized lens system with bloomed lenses, for two diameters of channel.

B. Breaking Strength and Flexibility

Although the theory describing the rupture mechanism in single fibres has been established, the same cannot be said for bundles of such fibres. In straight tension, the strength of a bundle of fibres is the combined strengths of the individual fibres comprising the bundle. However, if such a bundle is flexed round a mandrel then the fibres will be subjected to differing stresses, owing to the variations in bending radius with distance from the mandrel, and will also be subjected to abrasion from neighbouring fibres.

A satisfactory theory has not yet been developed for such a case, although it is obvious that the stronger the individual fibres the stronger the bundle. Figure 9 shows the results of increasing the load on a number of similar bundles which were bent through a right angle around a mandrel. The results for breaking strength were obtained by measuring the transmission of the bundle as the load was increased. When significant breakage occurred the effect was catastrophic since the strain on the unbroken fibres increased dramatically, so that a very large decrease in transmission occurred at this load, and it is this value of load which is recorded in the Figure. The desirability of lubrication is clearly seen from these results.

Unless extreme flexibility is required, the optical fibres are normally encased in a flexible trunking which protects the fibres and limits the bending radius to a value well above the critical value for fibre breakage. When such a light guide is flexed there is a decrease in transmission due to fibre breakage, but after a relatively small number of flexures this decrease stops, and the transmission remains constant at a value of

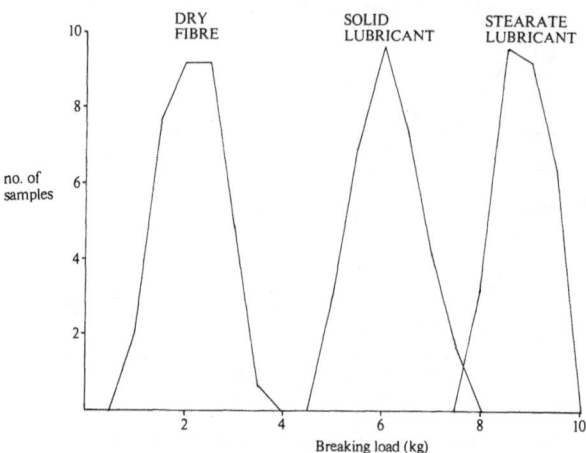

Fig. 9   Distribution of breaking strengths for unlubricated and lubricated fibre bundles bent at right angles around a mandrel 5.0 mm in diameter.

98–99% of the original transmission. This initial breakage is due to the presence of fibres in the light guide which were stressed highly during assembly or were weaker than the rest because of a scratched or damaged surface. Figure 10 shows the transmission of a 3 mm diameter light guide which was repeatedly flexed over a mandrel of radius 3 cm. The transmission was measured at intervals and noted; the initial drop in transmission can be clearly seen.

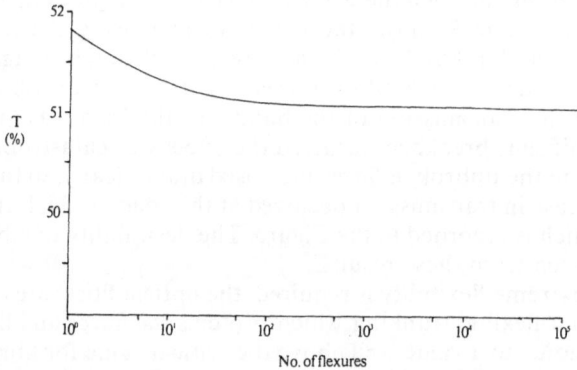

Fig. 10   Variation in transmission of a simple light guide subjected to flexure around a mandrel 6.0 cm in diameter.

C. Effects of Temperature

The glass components of the optical fibre can withstand an extremely large temperature range, although there is a decrease in transmission as the temperature increases. However, the operating range for light guides is determined by the temperature limits of the trunking and the adhesive used to bond the fibres. Typically, this gives working range of − 140°C to 250°C, although special guides have been made for operation up to 400°C.

IV. MECHANICAL PROPERTIES OF PLASTIC FIBRES

Plastic fibres are inherently more flexible than glass fibres and can be used as single fibres with diameters up to 2.0 mm in systems which require flexibility. This means that, for light guides with active diameters of 2.0 mm or less, plastic fibres have a higher transmission, since the core packing fraction is unity. Plastic fibres, however, are more easily damaged than glass and show a "fatigue" effect after repeated flexing. This is due to the crazing of the plastic at the core–sheath interface and seriously reduces the transmitting efficiency of these fibres.

Plastic fibres can be used continuously at 50°C and intermittently up to 90° C.

REFERENCE

1. J. Strong (Ed.), *Concepts of Classical Optics*, Freeman, San Francisco (1958), p. 553.

CHAPTER 5

Non-Coherent Bundles–Applications

The applications of non-coherent bundles of optical fibres constitute the major area of utilization of this technology at the present time, and this situation is likely to continue in the foreseeable future. This is because these are the simplest assemblies of optical fibres and therefore offer the greatest scope in variations of design and construction.

In the following sections the applications will be divided into four main groups, as follows:

1. Simple guides.
2. Beam-splitting guides.
3. Shape-changing guides.
4. Information displays utilizing complex bundles.

By definition the applications outlined in the last three groups will fulfil at least one of the required characteristics of a viable fibre optics system, as discussed earlier. Other characteristics may be fulfilled in certain of the applications, and these will be discussed in the appropriate section.

I. APPLICATIONS OF SIMPLE LIGHT GUIDES

The simple light guide consists of a bundle of flexible optical fibres (normally between 40 μm and 60 μm in diameter), each end of which is consolidated with a resin, ground and polished as described in section 4.II; the end cross-sections are of the same shape, normally circular Because of this definition, the characteristics of shape-changing and beam-splitting are not utilized, and viable applications must involve the other characteristics. The definition also means that the function of a simple light guide is to transfer light flux between two points. There are two modes in which this guide can operate: first, where light is to be carried to a remote point, which we will define as the transmission mode, and secondly, where light is to be received from a remote point to be used eventually to energize the eye or photo-detector, defined as the reception mode. The transmission mode is by far the more common, and the first application to be described operates in this mode.

A. Illumination for Medical Viewing Instruments

The basic medical viewing instrument consists of a long hollow probe which can be inserted into body cavities or wounds for examination purposes. In some cases, a telescope can be inserted into this probe and in others, the probe simply provides a line of sight to the region of interest for direct viewing. In both cases, illumination must be provided and the conventional method is to place a small filament bulb at the far end of the probe. There are two serious disadvantages to this method; the first is that the power consumption of the bulb must be low in order to prevent damage to the surrounding tissue, and hence the light output is low. The second disadvantage is that this method requires the passage of an electric current into the body of the patient which is extremely hazardous, particularly if the bulb fails.

The use of a simple light guide eliminates both of these disadvantages. In a fibre optics illumination system the source of light is external to the patient and is focussed on to the input of a simple light guide, the heat being removed from the beam by the use of heat filters or dichroic reflectors. The guide is positioned in the probe so that the output beam illuminates the required area. A diagrammatic sketch of a practical system is shown in Fig. 1, in which it will be noted that two light guides are used in series. The first is a 2 m long flexible guide which takes the light from the source to the instrument enabling the light source to be placed at a convenient distance from the patient. This is coupled to a second guide which carries the light along the instrument to the distal end. This second guide is normally not flexible, but is encased in a nickel–steel tube to facilitate insertion and removal. This tube has the same diameter as that used to contain the bulb leads in

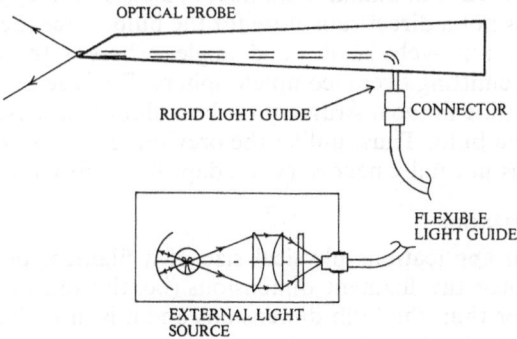

Fig. 1   Provision of illumination from an external source to an optical probe for medical inspection.

the conventional system, so that these can be interchanged without modification.

The resulting illumination is much higher than the conventional lighting system, so much so that the introduction of fibre optics into this field has greatly increased the diagnostic capabilities of an internal examination. In addition, the use of cine photography is now feasible in certain examinations, which was hitherto impossible. In more sophisticated systems, provision is made for the alteration of the colour of the light and for flash photography. As will be discussed in a later chapter, there is a range of flexible viewing instruments where the image is carried by a fibre bundle, and in these a light guide is an obvious choice for the provision of illumination.

This application is an ideal utilization of the transfer properties of a light guide; the source of illumination is removed from its point of application, where it could be hazardous, to a safe position.

B. Instrument Illumination in General

In situations where instruments may be used in a dim light, the instrument dials are provided with some illuminating means. This normally consists of a filament bulb, or bulbs, positioned within the instrument case in such a way as to provide the necessary degree of illumination. Where a number of such instruments are mounted in a panel, for example in motor cars or aircraft, the replacing of one of these bulbs can be a major undertaking, since the entire panel may have to be removed to gain access.

This is another ideal application for moving the source to a remote point, in this case to ease replacement, and guiding the light via a light guide to the point of application. An additional benefit is that a single bulb can be used to illuminate a number of dials. In this application, the light guide is not a direct substitute for the bulb, since the guide emits light over a relatively small solid angle, whereas the bulb can be regarded as emitting over a complete sphere. Because of this, the position of the guide in the instrument for best illumination is not the same as that of the bulb. Thus, unlike the previous case, a re-design of the instrument is normally necessary to adapt it for fibre optics.

C. Miniaturization

In certain applications physical size of a filament bulb can pose problems since the filament dimensions (i.e. the source of light) are much smaller than the bulb dimensions and it is this latter size which determines the packing density achievable. Thus, where miniaturization is essential the use of fibre optics can prove to be advantageous. An obvious case in mimic diagrams where the use of light guides enables a better design to be achieved, in addition to solving maintenance problems and reducing the amount of heat dissipated on the actual panel. In a similar way, simple light guides have found extensive application in flight simulators, where the scaling down of lamps can be a serious problem. Figure 2 shows the layout of a typical simulator, in which all the lights were simulated using light guides, some with a diameter of only 0.5 mm, to give a perfect scaled-down version of the actual view.

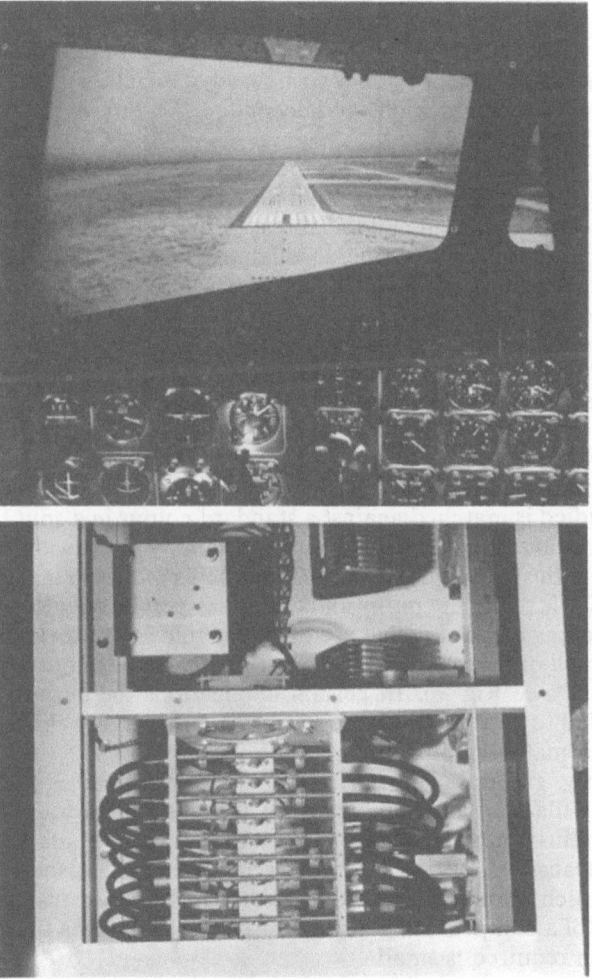

Fig. 2   The use of light guides in flight simulation:
(a) The pilot's crew of a simulated landing strip in which light guides are used to represent the lighting systems.
(b) Banks of light guides and light sources inside a flight simulator. (Courtesy of Redifon Flight Simulation Ltd)

D. Switching

Use has been made of the insulating properties of glass optical fibres in electronic switching applications where the light guide is used to couple two circuits without introducing a direct electrical link. In the primary circuit, an L.E.D. (light-emitting diode) forms the load so that a light output is obtained from this which is proportional to the current in that circuit. This light is collected by a simple light guide and is transmitted to a photo-detector in the secondary circuit, which provides an electrical signal in this circuit corresponding to that in the first. This type of link provides two main advantages: first, the electrical isolation can be very high, and secondly, the link is immune from electrical interference.

One application of this link is the coupling of integrated circuits so that interference is kept to a minimum. In another, the L.E.D. is in a high-voltage line (about 200 KV above ground) and the light output is a measure of the current flowing through this line. This light is collected by the light guide and led to a safe area where the emerging beam is measured by a photo-detection system which is calibrated to indicate the current in the line. An expensive isolation transformer is not required, and no interference arises from the large voltage gradients which exist close to the line.

E. Colorimetry

In colour matching, use is made of an instrument whereby the colour being studied is matched against a standard colour in a split field. The standard colour is generated from white light by interposing a number of filters in the standard beam, the filters being mounted in turrets. The instrument is relatively bulky and the optics permit only the matching of large areas of colour. The use of a light guide to pick up the colour from a small or inaccessible area greatly extends the range of use of the instrument (see Fig. 3). In certain applications the output from the guide is fed to a single fibre-integrating chamber to provide a uniform colour for matching purposes.

Similarly, guides have been used coupled to pyrometers, of the disappearing-filament type, to measure furnace temperatures in steel works. In this type of application allowance must be made for the fact that the transmission of the guide is not neutral with respect to wavelength, which normally means a re-calibration of the instrument or the insertion of a compensating filter; however in both cases the degree of correction required is small.

F. Visual Monitoring

A great deal of interest has been shown in the use of simple light guides for monitoring lamps or flames where these are not in a suitable

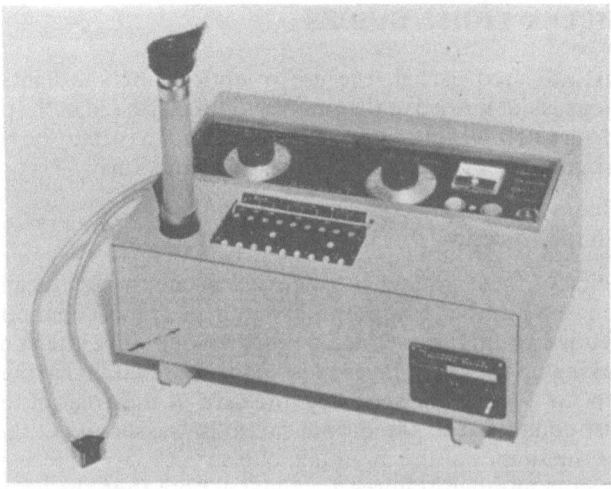

Fig. 3 Photograph of a colorimeter, incorporating a flexible fibre optics probe. (Courtesy of The Tintometer Ltd)

position for direct vision. The most importance example of this is the monitoring of car lights by the driver. In this application the input end of a simple light guide is positioned to accept light from each lamp, i.e. side lights, brake lights, and direction indicators. The output ends of these guides are positioned in a mimic diagram of the car lighting system fixed to the facia panel. The driver can then check the state of his lights from the driving position. The advantage of this system is that it is passive and completely reliable.

One disadvantage is that failure of a bulb eliminates only one illuminated point from the mimic diagram which would not be immediately obvious to the driver. This is an inevitable consequence of the system, but can be avoided by the following technique. The light from the lamps is filtered to give a greeny-blue light and the output from each guide is mixed with red light coming from a subsidiary bulb within the monitoring unit. The intensities of both beams are adjusted so that the first colour swamps the second. Thus when all lamps are functioning each point in the mimic diagram emits a greenish light, but when one lamp fails the corresponding point in the mimic diagram emits a red light, which is immediately discernible by the driver.

In another application a light guide is used to view a pilot light in a gas-fired boiler. In this case a small piece of ceramic material is heated to incandescence by the pilot flame to give a visible indication of its presence. The light guide transfers the light to a point where it can be conveniently seen.

II. COMPLEX LIGHT GUIDES

As was indicated earlier, the use of optical fibres in light-guiding applications is not limited to the simple guides discussed in the previous section. More complex designs are possible, and these may be regarded as performing one or both of the following functions:

1. Beam splitting or combining.
2. Shape changing.

In the first of these, branched guides are used in which the fibres are split amongst a number of inputs and outputs, the most common being a single-ended to multiple-ended configuration. Since the sum of the fibres making up the input faces must equal that making up the output faces then, to a sufficient accuracy, the sum of the areas of the input faces must equal that of the output faces. In the second of the above functions the shapes of the input and output faces are different. However, since the number of fibres are equal in both then the areas of the input and output faces are equal.

The manufacturing techniques for this type of guide are similar to those for simple guides, although the design of terminations is necessarily more complex.

III. BRANCHED LIGHT GUIDES

A. 'Y'-Guide Applications

The simplest form of branched light guide is one in which the guide has two branches, and for obvious reasons this has become known as Y-guide. This can be used to combine two beams of light, for example in illumination applications, where it is necessary to provide a reserve light source in the event of failure of the main source. However, the most common use of this type of guide has been in sensing applications, and the normal mode of operation is illustrated in Fig. 4 where it will be seen that one branch is used to illuminate the area of interest, and the other to guide the reflected light to a photo-detector. Any change in the amount of reflected light will result in a change of signal level from the detector, which can be used to operate a control circuit.

The theory of operation of such a system is best illustrated by considering a Y-guide of the co-axial type, where the emitting guide is central. Such a guide is shown in Fig. 5 where the radius of the emitter guide is r and of the total bundle is R. This guide is positioned normally to a reflecting surface and a distance d from it.

It can be shown (1) that the fraction of light intercepted by a disc of

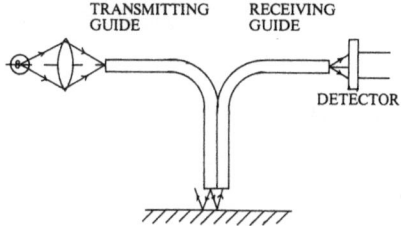

Fig. 4 Basic diagram of a Y-guide operating as an opto-electronic sensor.

radius r_2, from a parallel disc of radius r_1 emitting light in a Lambertian fashion is given by:

$$F(r_1, r_2, l) = \frac{1}{2r_1^2} \left\{ r_1^2 + r_2^2 + l^2 - \sqrt{(r_1^2 + r_2^2 + l^2)^2 - 4r_1^2r_2^2} \right\} \quad (1)$$

where l is the separation of the discs. From Fig. (5) it can therefore be seen that the fraction of light intercepted by the receiving guide will be given by:

$$F_1 = \rho(F(r, R, 2d) - F(r, r, 2d)) \quad (2)$$

where ρ is the reflectivity of the surface, if light is emitted from the central guide in a Lambertian fashion. This can be written as:

$$F_1 = \frac{\rho}{2r^2} \left\{ R^2 - r^2 + \sqrt{16d^4 + 16r^2d^2} - \sqrt{(r^2 + R^2 + 4d^2)^2 - 4r^2R^2} \right\} (3)$$

The variation of this fraction with separation is shown in Fig 6 for the case where the two guides have equal areas, i.e.

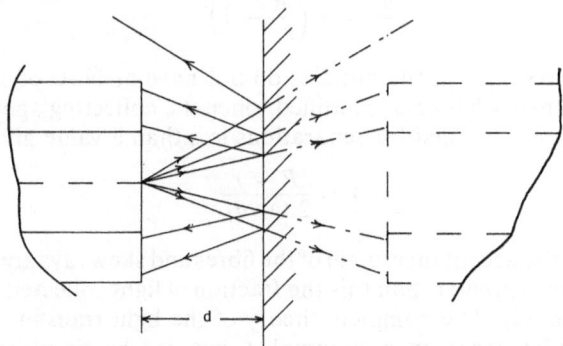

Fig. 5 Diagram showing paths of typical rays from a coaxial Y-guide, defined by radii r and R ($r < R$), being reflected from a surface at a distance, d, from the output face.

Fibre Optics

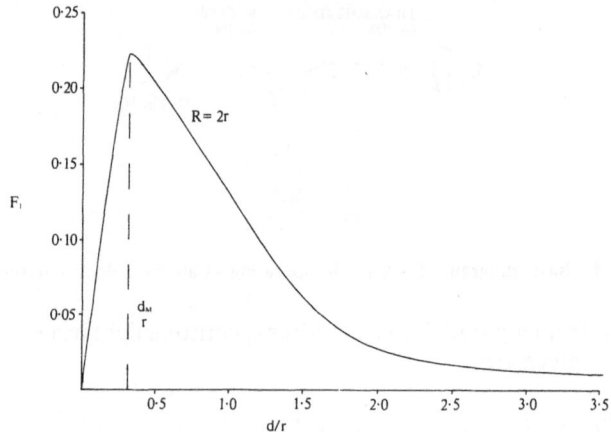

Fig. 6 Variation of the fraction of light (F_1), collected by a coaxial Y-guide ($R = \sqrt{2}\, r$) plotted as a function of the separation of the reflecting surface $\dfrac{d}{r}$, ($\rho = 1.0$).

$R = \sqrt{2}\, r$ and $\rho = 1.0$. It will be seen that there is a region below the maximum where the curve is almost rectilinear so that in this region the system behaves as a linear transducer.

By differentiating Eq. (3) and setting this equal to zero, the condition for maximum light interception is found to be given by:

$$d_M = \tfrac{1}{2}(rR - r^2)^{\frac{1}{2}}$$

or

$$\frac{d_M}{r} = \tfrac{1}{2}\left(\frac{R}{r} - 1\right)^{\frac{1}{2}} \tag{4}$$

If the fibres used in the bundles do not have an N.A. of unity then the above theory has to be modified, since the collecting aperture will not be completely filled for separations less than a value given by:

$$d = \frac{R + r}{2 \tan \theta_M} \tag{5}$$

where θ_M is the acceptance angle of the fibres and skew rays are ignored. At separations greater than this the fraction of light collected will be as given by Eq. (3). The complete theory of the light transfer with restricted angular acceptance is complex, but can be simplified if it is assumed that $r \ll R$. In this case, the fraction of light collected is given by replacing R in Eq. (3) by the variable expression $(2d \tan \theta - r)$ which takes account of the incomplete coverage of the collecting bundle by the output. This formula is applicable up to a separation given by:

$$d = \frac{R + r}{2 \tan \theta}$$

Above this the collected light is given by Eq. (3). The effect of the N.A. of the fibres can be seen in Fig. 7 where the fraction of light collected for $R = 10r$ is plotted for N.A.s of 1.0 and 0.5.

Detection of Reflectivity Changes.

The simplest application of this type of guide is to be found in sensing devices where the separation of the guide from the surface is constant, and the change in light level is produced by a change in reflectivity of the surface. The cause of this change in reflection will vary according to the specific application, and a number of the more important ones will now be discussed.

A non-contact counting device has been designed using a Y-guide in which the desired change is caused by the presence or absence of a reflecting surface, at or close to the common end of the guide. This surface need not be specularly reflecting, and most substances will reflect back enough light to operate the device so that, in most cases, one can use the surface of the objects to be counted. The actual reflectivity of the surface will determine the maximum operating distance

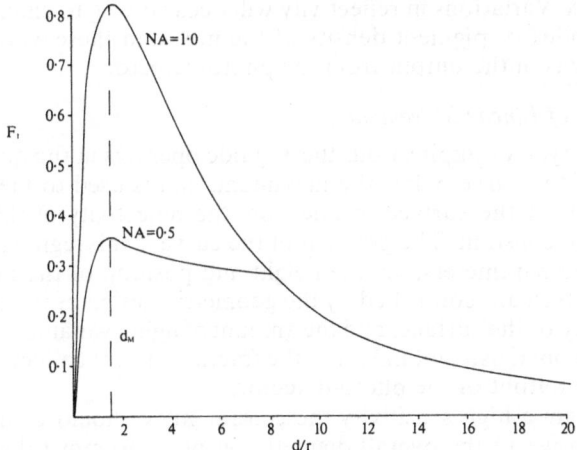

Fig. 7   Variation of the fraction of light (F_1) collected by a coaxial Y-guide ($R = 10\,r$) plotted as a function of the separation of the reflecting surface $\dfrac{d}{r}$ for optical fibres with N.A. = 0.5 and 1.0 ($\rho = 1.0$).

from the end of the guide to the object, which can be up to 30 cm for a
high reflectivity, e.g. metallic surfaces. This approach has also been
used to provide automatic positioning means, particularly for steel
strip and printing paper. In this, the desired lateral position of the
strip in a piece of equipment is established by suitably positioning two
Y-guide sensors the width of the strip apart, at right angles to the
motion of the strip. If the strip is correctly positioned, both sensors
will receive reflected light back from the strip; if not, only one sensor
will receive this light and the appropriate lateral correction can be
made. The sensitivity of this device is dependent upon the diameter of
the guide used and the sensitivity of the detection system, but positional
accuracies of better than 1.0 mm can readily be achieved.

In mark-sensing applications, the reflectivity change is caused by a
mark on the surface being examined. This surface is usually paper,
which has a specular component in its reflection characteristics, and
this can mask an imprinted mark if the specular component is sufficiently large. The effect of this can be effectively eliminated by placing
the common end of the Y-guide almost in contact with the paper so that
only scattered light can reach the detector. The main area of application for this type of device is in the analysis of stock records or question-
naires, which can be suitably designed to be completed on a "mark-
no-mark" basis.

This principle has also been used in the paint and confectionery industries to test the homogeneity of mixtures, by immersing the Y-guide
in the mix. Variations in reflectivity will occur owing to changes in refractive index or pigment density of the mix, and these will appear as
fluctuations in the output from the photo-detector.

Detection of Linear Movement.

In this type of application, the Y-guide operates in the quasi-linear
region of the curve below the maximum, and is used to measure the
movement of the surface in question, the reflectivity of this surface
being kept constant. The gradient of the curve in this region is controlled by two parameters, viz the height and position of the maximum.
These in turn are controlled by the geometric design of the guide, the
reflectivity of the surface, and the amount of light available. The last of
these will obviously not influence the fraction of light collected but will
affect the output of the photo-detector.

To obtain a high sensitivity the central guide should be made very
small relative to the overall diameter, in order to move the position
of the maximum close to the origin. However, this will correspondingly
reduce the amount of light fed into the system, which will reduce the
maximum output from the photo-detector tending to counteract the
effect of moving the maximum. To overcome this, the illuminating

fibres are split into a number of small outputs uniformly distributed throughout the bundle. Each output can be treated according to the above theory to a sufficient accuracy, each sharing a common collecting bundle. However, the fraction of light intercepted for any one output will have to be reduced by the ratio of collecting area to total bundle area outside that output, since the light intercepted by the other outputs will obviously not reach the detector. As an example of this, the maximum sensitivity is achieved if every other fibre in the common end is a source so that the value for r is the fibre radius. The areas of source and collector are equal so that maximum light transfer is possible. Fig. 8 shows the theoretical output curves obtained for a guide with an overall diameter of 2 mm, using 50 μm fibres as described above, and with equal emitting and collecting areas.

The above system has obvious attractions as a linear transducer. This are:

1. It is a non-contact device.
2. The sensor tip can be made very small (a fraction of a millimetre, if required).
3. The frequency response is high, limited only by the detector response.
4. The electrical circuitry can be placed some distance from the point of measurement so that electrical interference between neighbouring transducers can be eliminated.

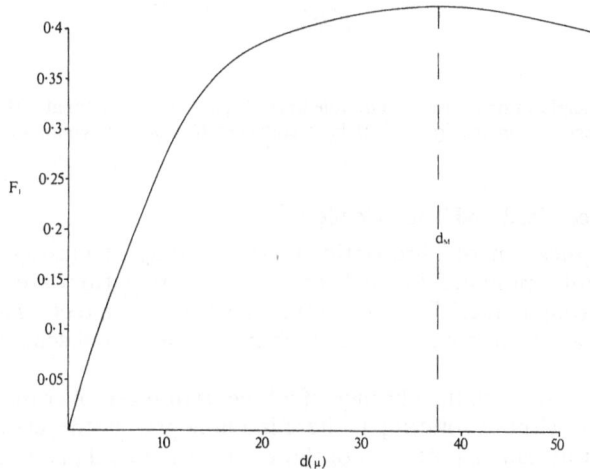

Fig. 8 Variation of the fraction of light (F_1) collected by a multiple aperture Y-guide plotted as a function of the separation of the reflecting surface. The aperture size equals the fibre size (50μ) and the probe diameter is 1.0 mm ($\rho = 1.0$).

Because of these advantages this device has found application in multi-gauging, where the dimensional accuracy of a component is checked by placing it in a jig, containing a number of suitably positioned transducers, so that all the relevant dimensions can be monitored simultaneously. This is of particular value where the components have a complex shape, e. g. turbine blades. Simple devices of this nature can have sensitivities of a few hundred $\mu V/\mu m$.

The major disadvantage of this device is the variation of sensitivity with surface reflectivity. This can be overcome by using two of these devices in parallel, operating at the same point but at slightly different separations, as shown in Fig. 9. If this difference in separation is known then the difference in outputs, which correspond to a known separation, can be used to compensate for changes in reflectivity. Another approach is to convert the device into a contact transducer by incorporating a spring-loaded stylus which moves a reflector relative to the probe, as shown in Fig. 10. By this means, the reflectivity can be kept constant, although at the expense of losing the non-contact nature of the device and of a reduction in frequency response.

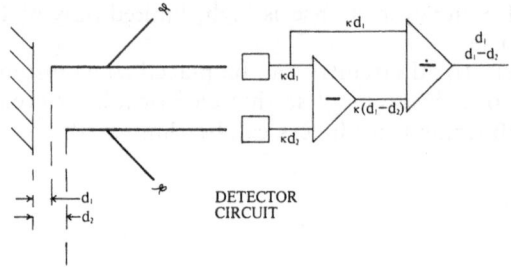

Fig. 9 Principle of operation of a double beam Y-guide which eliminates the effect of surface reflectivity on the fraction of light collected. ($k = \rho \times A$, where A is the guide sensitivity).

B. Punched-Card and Tape Readers

The application of fibre optics to the reading of messages in the form of holes punched in cards or tape is a straightforward use of a beam-splitting guide.[2] The conventional method of detecting a hole in a card is an electro-mechanical technique, illustrated schematically in Fig. 11.

The star wheel, in the absence of a hole, is dragged over the surface of the card without rotating. If there is a hole present the leading spur falls into this causing rotation of the wheel as the card passes and this is detected electrically. There is one of these "feelers" for each row on the card. Since this detection is mechanical the reading speed is slow and problems are encountered owing to mechanical wear and the creation of dust from the paper.

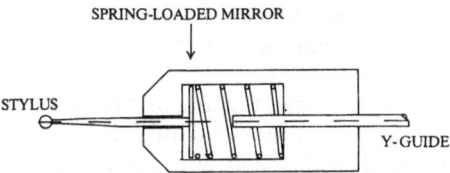

Fig. 10   Diagram of a contact linear transducer using fibre optics.

An electro-optical approach would solve these problems and the obvious method is to illuminate one side of the card or tape and detect the presence of holes by an array of photo-detectors on the other side, there being one photo-detector for each possible row. There are three possible methods of providing the illumination for this. The most obvious is to use a lamp whose filament is focussed on to the card and, to ensure a uniform distribution of light, a ribbon filament is used. In practice, this system proves unsatisfactory for the following reasons.

First, the area which requires illumination has a large aspect ratio, up to 100:1 for cards. This is much greater than the aspect ratio of a ribbon filament which may be typically 5:1, which means that if the height of the illuminated area is correct, the width will be too large by a factor of 20. Since this could cause false indication of holes which were not in the column being read, a mask of the correct width must be placed at the surface of the card or tape. A more serious consequence of this is that, to provide a given illumination at a hole position, a lamp must be used which has up to 20 times the power ideally required. In fact the situation is much worse than this since allowances must be made for tolerances in filament position and the sag of the filament during the life of the lamp. As a result, the lamp power required gives rise to heating problems, the lamp must be fan-cooled and the system becomes bulky and difficult to design into the equipment.

A second approach is to use a number of small bulbs, there being one bulb for each possible row of holes. This system overcomes the power problem of the first, since only a small area around each possible hole

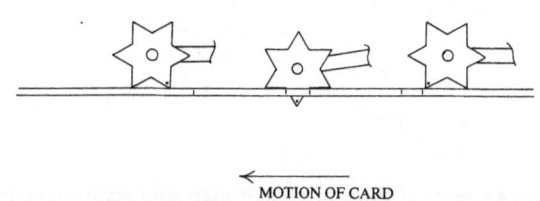

MOTION OF CARD

Fig. 11   Principle of operation of mechanical detection of holes punched cards.

need be illuminated. Care must be taken in the design of the mounting block, since the bulbs emit light in all directions, to prevent cross-talk between adjacent detectors. However, this system suffers from two main disadvantages. First, the loss of a bulb would cause misreading since this would only affect one hole position and secondly, the output from the small bulbs that are required can vary enormously throughout their life, and the rate of change is not constant for a given design of bulb. Therefore even if these bulbs were initially selected to give equal illumination this cannot be maintained for any length of time, and the imbalance creates insuperable problems in the setting of trigger levels in the detection circuitry.

The third method, using fibre optics, combines the advantages of the first two with none of the disadvantages. The bundle used is a beam-splitter, which divides the output of a lamp into a number of outputs equal to the desired number of rows. These outputs are then positioned opposite possible holes in the cards or tape, as shown diagrammatically in Fig. 12.

Since the area requiring illumination is much smaller than that in the method first discussed, the lamp can be of low wattage and does not require forced cooling. It may help in appreciating the significance of this to consider the following example.

It is found that the output from a fibre bundle 1.0 mm in diameter is amply sufficient to trigger the photo-detectors used. Thus, for a 12 bit column, a beam-splitter having 12 outputs, 1.0 mm in diameter, and an input just over 3 mm in diameter is used. The column may be typically 100 mm in height so that using the first method one must illuminate an area of 100 mm × 20 mm, if a filament with a 5:1 aspect ratio is used. Thus the ratio of illuminated area is about 200:1. The N.A. of the fibres used is chosen so that no light from one output can reach a hole in an adjacent column; cross-talk is therefore negligible with the fibre optics

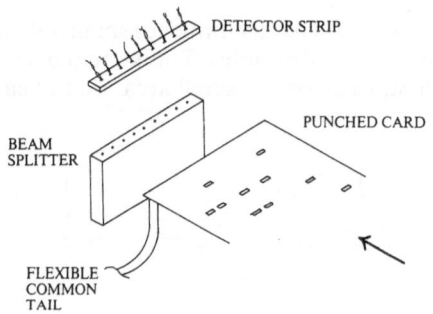

Fig. 12   Schematic representation of a non-contact card reader using fibre optics.

system. Since only one lamp is used, the failure of this stops the illumination of all outputs and this can immediately be detected. A further feature of the fibre optics system is that the lamp can be remote from the reading station, eliminating heat at the reading station and easing maintenance.

It can be seen, therefore, that in this particular application the use of fibre optics solves all the problems and gives two additional advantages. It is because of this that this type of application was the first commercial use of fibre optics on a large scale.

C. Electronic Switching

This type of application utilizes the high electrical insulation of a bundle of optical fibres, which means that circuits which differ greatly in electrical potential can be directly coupled by optical means. A typical case is to be found in high-voltage rectifiers where the intention is to switch a chain of high-voltage SCR's which are being used to rectify a 300 KV line. Each rectifier is switched by a signal derived from a photo-detector which can be illuminated by light from one output of a branched guide. This guide has a single input which can be illuminated by a solid state light source. The system is shown schematically in Fig. 13.

In practice, the design is more complex in order to incorporate safety features, and the final system has three of the above systems assembled in parallel, each fed with its own source. In this, therefore, each output comprises three bundles, each from one of the light sources; if one source fails the other two are still sufficient to trigger the detectors. In addition, an extra output is included in each beamsplitter which is fed to a monitoring device that will give an indication of source failure.

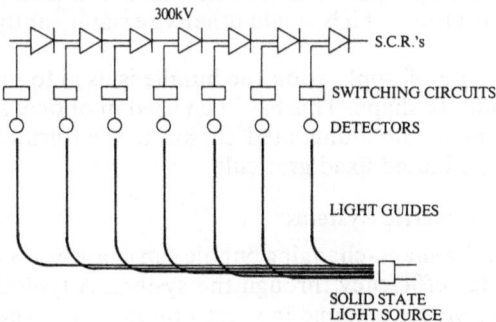

Fig. 13 A typical circuit for the switching of high voltages using fibre optics. The electrical isolation of the primary and secondary circuits is provided by the high dielectric strength of the glass fibres.

IV. SHAPE-CHANGING SYSTEMS

The ability to change the cross-section of a beam of light using fibre optics arises from the fact that the individual fibres in a bundle act independently, so that they need not be related spatially in any way. The only design criterion is that the areas of the end cross-section be equal. A shape-changing bundle, as these are called, normally forms an interface between an optical system, which acts most efficiently on beams of circular cross-section, and some other system, so that one end of the bundle is normally circular.

A. Uses in Illuminating Systems

A shape-changing bundle can be used to improve the efficiency of light transfer to an area requiring illumination, and would be inserted either in the condensing system or at its output. In the former position the bundle transforms the image of the lamp filament into a circle to allow a more efficient utilization of the subsequent condensing optics. In this application the introduction of fibre optics also provides the opportunity to fold the condensing system or to remove the lamp to a more accessible position to ease replacement. Since the input of the bundle operates in a high-temperature environment, a high-temperature resin is used to bond the fibres, and this end is not ferruled. In the latter position, the bundle transforms the beam cross-section to match that of the area to be illuminated. An example of this is to be found in document scanning, in which the area to be illuminated is a line of print, where the output of the bundle would be in form of a rectangle whose width equalled that of the document. If the fibre bundle is flexible then the scanning head can be moved without the necessity to move the light condensing system, which greatly reduces the weight to be moved. An additional advantage gained is that the lamp can then be free from mechanical vibration, which would otherwise significantly decrease its life.

In another type of application the bundle is used to provide a light source of a definite shape. This has been used in optical sights to provide either a moveable illuminated cursor, in the form of a cross or a 'vee', or an illuminated fixed graticule.

B. Uses in Photometric Systems

In this area, the shape-changing bundle can again be used to increase the light-transfer efficiency through the system. A typical example of the application of this is found in spectrophotometry where the spectrograph input is a fine slit. Obviously the illumination of this slit by conventional optics is very wasteful of light, and a more efficient system results from the use of a fibre system in which the circular cross-section of the input is transformed into a rectangular area, whose dimensions equal that of the input slit. An application of this nature in stellar photometry has been described by Kapany.[3] This 'slit-to-circle' design of bundle has also been used to sample a fine rectangular area within a fringe pattern and transfer the light efficiently to a photo-detector. In this application a number of such bundles were mounted with their rectangular faces stacked together so that adjacent areas in the pattern could be measured simultaneously, whilst the outputs were spaced to allow each bundle to feed a photo-multiplier (Fig. 14). Alternatively, a single flexible bundle attached to a mechanical scanning head could have been used to scan the entire pattern, giving a much larger sampling area at the expense of simultaneity of measurement. If a multiple head is to be used then each bundle must be calibrated, since the individual transmissions of the bundles will vary. A coaxial version of this device has been described by Forrest.[4]

C. The Collimation of the Output of Semi-Conductor Lasers

The design of a collimating system for semi-conductor lasers can be made much more compact by the use of a shape-changing fibre bundle. The emitting area in a semi-conductor laser is a narrow rectangle approximately 1.0 mm $\times$ 5.0 μm and diffraction spreads the emergent beam into a fairly wide angle in a direction perpendicular to the larger dimension of the emitting area, with a semi-angle of around 20°. For most applications a collimated beam is required and the laser is normally operated at the focal point of a lens system. The divergence of the resultant beam is determined by the subtense of the source at the principal plane of the lens, which means that the degree of collimation is determined by the larger dimension of the source. It is obvious that in the case of a rectangular source with a large aspect ratio the focal length required for a given divergence will be larger than if the source were square, or circular, with the same active area. For a source with

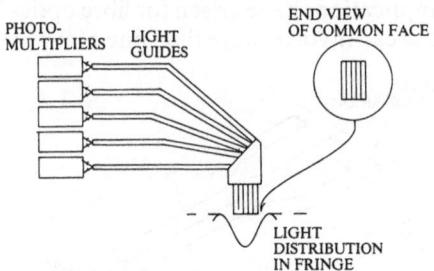

Fig. 14 Diagrammatic representation of the dissection of an interference fringe using closely packed shape changers.

the dimensions mentioned above the focal length required is some 70 times as great as that for a square source of the same area for the same degree of collimation. To attain collimation to a few milli-radians focal lengths of 30 to 50 cm are required with lens diameters in the range of 20 to 30 cm. Since the attraction of the semi-conductor laser lies in its compactness, these values are inconveniently large.

A significant improvement is achieved by using a square- or circle-to-rectangular converter made with optical fibres. The rectangular input face must necessarily be larger than the emitting area to allow for positioning tolerances, so that the practical reduction in focal length is not as great as is theoretically possible. In a typical bundle, the input face would consist of a single layer of 20 μm fibres, 1.0 mm wide grouped into a square or circular format at the output end. This gives a reduction in the required focal length by a factor of $\times$ 7, which means that the required focal length is in the range 4–7 cm, with lens diameters of 3–6 cm, which allows a compact design to be achieved.

An added advantage gained through the use of fibre optics is that the outputs from a number of semi-conductor lasers can be combined into a single output by the use of a branched guide, in which the input face of each branch is rectangular and the outputs are combined into a single square or circular output (Fig. 15). This enables higher output powers to be achieved than is possible with a single laser and, for example, makes the design of a hand-held laser rangefinder a feasible proposition. Present-day semi-conductor lasers have to be operated at low temperatures to give an acceptable conversion efficiency and are normally mounted on a cooled metal strip. To prevent serious heat losses from this, the optical fibres are mounted in a non-metallic material, e.g. epoxy resin.

V. LIGHT GUIDES IN DISPLAY SYSTEMS

A number of applications have arisen for fibre optics in the display of information. These can involve more than one of the basic types of light

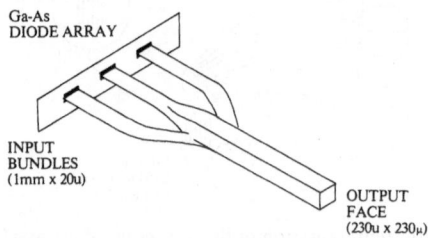

Fig. 15   Diagram of a beam-splitter used to combine the outputs from three Ga-As laser diodes.

guide, or a combination of light guides and single optical fibres. A number of such systems will be discussed in this section.

The ability of fibre optics bundles to alter the shape and distribution of a beam of light has a number of uses in displays of information. One such use, in mimic diagrams, has already been outlined in section 5.I.C. The basic principle behind all display applications is to use a complex light guide, the input of which is designed to receive the maximum amount of light from a condensing system, and the output of which is designed to display the required character or legend. Thus a single character or legend may be displayed as indicated in Fig. 16. However, when it is necessary to be able to display a number of items in the same area, problems can arise from the overlapping of these items.

A. Digital Displays

One way of avoiding this difficulty is to build up the desired display using a number of separate light units which can be shared amongst a number of characters. This is typified by the 7-bar numeric system which is illustrated in Fig. 17. In this, seven separate light units are employed and the digits from 0 to 9 can be displayed, albeit in a stylized fashion, by switching the appropriate light units. (Figure 18) By introducing diagonal units, and additional vertical units through the centre of the display, a full alpha-numeric capability is provided, requiring a total of 14 light units. These displays are available using moulded plastic units to form the bars, each being illuminated by a filament lamp, but they have a low output and suffer from uneven illumination within each bar. The use of fibre optics enables the light from the lamp to be more efficiently transformed into the required shape with a significant improvement in the appearance of the display. A further improvement is achieved by forming each bar from a series of circular outputs as shown in Fig. 19 which softens the angular appearance of certain digits and gives a more pleasing display. Such displays have typically a brightness of 1000 foot-lambert which renders them visible

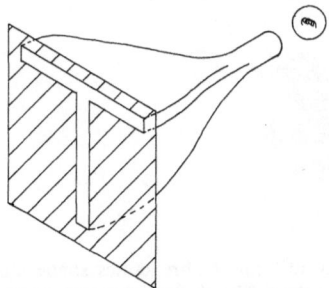

Fig. 16   A character display using a shape changing fibre bundle.

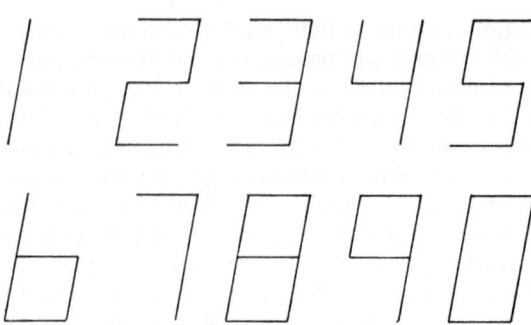

Fig. 17   The digits 0–9 represented on a 7-bar display system.

in high ambient lighting. Recently this type of bundle has been used in conjunction with L.E.D.s, where the fibre assembly is embedded in an opaque plastic to prevent the back-reflection of ambient light from the metallic connectors of the L.E.D. package. This reflected light can seriously degrade the readability of such displays.

Another approach is to form the digits from a matrix of light-emitting apertures, which can be preferentially energized (Fig. 20). If the digit can be sensibly constructed by apertures which are not adjacent it is possible to have overlapping displays where each aperture is utilized by only one display, in which case each legend can be treated separately as described previously. This is illustrated for the

Fig. 18   A digital display utilising 7 fibre optics shape changers, using the system illustrated in Fig. 17. (Courtesy of Rank Precision Industries Ltd)

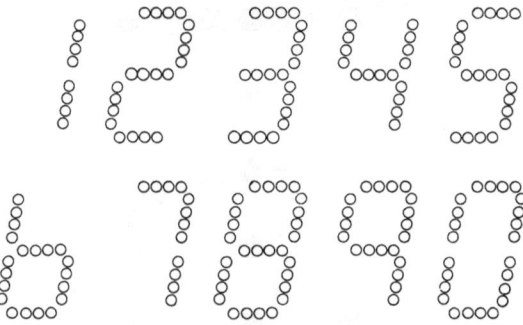

Fig. 19   The digits 0–9 represented on an improved 7-bar display system.

digits 1, 2 and 3 in Fig. 21. It is apparent that the larger the number of digits, the further apart must be the apertures forming each digit. For example, the digits 0 to 9 require an aperture spacing of around 6 aperture diameters to avoid the sharing of apertures. This is only marginally acceptable and such an approach is normally restricted to cases where only a maximum of about four legends is required.

B. Multi-Legend Displays

If, as is normally the case, a larger "vocabulary" is required, then the design must permit the sharing of apertures among a number of characters. One approach to this will be outlined for the case of warning signs for motorways.

These warning signs are required to warn motorists on fast roads of a number of hazards which may lie ahead or to recommend a reduced speed limit in dangerous driving conditions. The legends are constructed from an 11 × 13 matrix of 50 mm apertures, each of which can emit a powerful beam of light within ± 5° of the aperture axis.

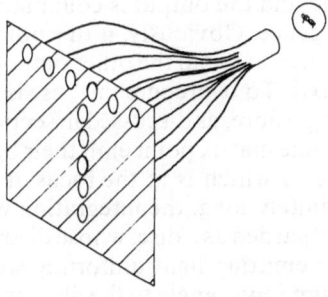

Fig. 20   A character display based on a matrix of light-emitting apertures.

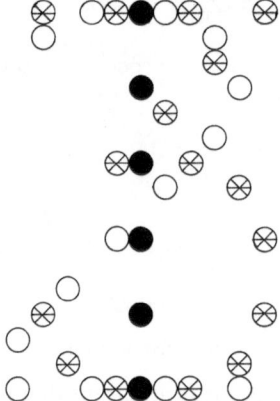

Fig. 21   The digits 1, 2 and 3 displayed on a matrix of light-emitting apertures, which are not shared.

The aperture brightness is in excess of 2500 foot-lambert and the legends are visible in bright sunlight at a distance of 400 m from the sign. In the conventional design there is a filament bulb at each matrix point, situated at the focus of a parabolic reflector. The desired legend is displayed by switching the appropriate lamps via a diode logic matrix. The maximum number of legends was sixteen so that sharing of matrix points was inevitable and, in fact, up to ten legends might have to share a particular point. The main disadvantages of this design were that lamp failures were difficult to monitor, and the display of different coloured legends was impossible. The introduction of fibre optics into the design eliminated these.

The basic design concept is to focus the light from a powerful filament lamp on to the input of a branched light guide which provides the illumination for one legend. Each branch is led to a matrix point required in the legend and the output is collimated by an aspheric lens to the required divergence. Obviously, if this point is shared, only one output will be on the lens axis and the other outputs will be asymmetric with respect to this axis. To overcome this, use is made of the integrating properties of a single fibre, as discussed in section 3.IV.E, and all the branches which feed one matrix point emit their light into a large single fibre, the output face of which is at the focus of the collimating lens. If this fibre were infinitely long, the integration would be perfect and the output could be regarded as a disc, whose diameter equalled that of the single fibre core, emitting light uniformly within an angle determined by the maximum input angle to the fibre or the N.A. of the fibre, whichever was the lower. However, in the practical design the single

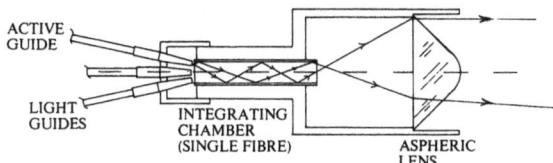

Fig. 22 The basic optical system at each matrix point in the motorway warning sign.

fibre is 6 mm to 10 mm in diameter and the length is restricted by the dimensions of the sign, to about 8 cm. The final design of the optical system at each matrix point is shown in Fig. 22. The N.A. of the fibre integrating chamber is 0.5 and is filled by the output from the beam splitters; this means that a maximum of only four reflections is possible and the integration is therefore far from perfect. The light output could, in fact, be represented as a series of annular sources surrounding a disc source, the beam-splitter output, situated at the input of the single fibre, which emitted light through an aperture equal in diameter to the single fibre core positioned at the output face of the single fibre. This is illustrated in Fig. 23 where it will be seen that each annulus corresponds to a specific number of reflections. The maximum divergence of the light issuing from the lens is determined by the subtense of the fibre core at the lens, but the light output is not uniformly distributed over the resulting solid angle. This is corrected by placing a lenticular plastic screen over the lens to spread the emergent light beam.

To increase the azimuthal uniformity still further, a small fibre plate

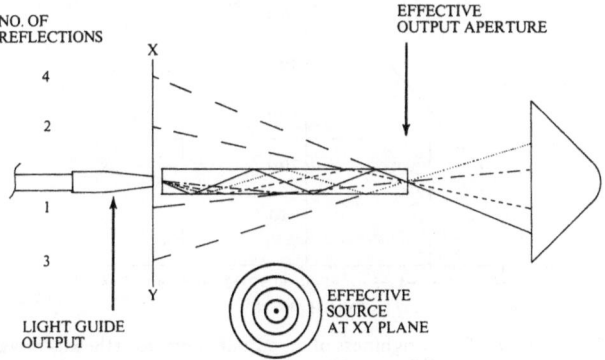

Fig. 23 The derivation of an effective source shape for the system of Fig. 22.

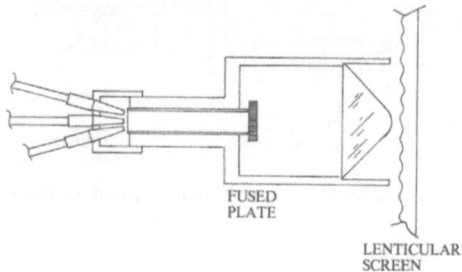

FUSED
PLATE

LENTICULAR
SCREEN

Fig. 24 The final design of the optical system at each matrix point in the motorway
warning sign.

is attached to the output face of the single fibre. The fibre size ($50\,\mu$m)
in this plate is very much less than the plate thickness (2.0 mm), to
permit good integration. The final design is illustrated in Fig. 24 in
which the output is still asymmetric but satisfies the specification as
shown in Fig. 25. A production version of this sign is shown in Fig. 26
displaying two out of sixteen possible legends.

In this design of sign, a coloured legend can be produced by simply
inserting a filter in the condensing system of the appropriate lamp.
Lamp failure can be detected by monitoring the current drawn by the
sign or by photo-electronic means. This latter method can also be used
to energize a reserve lamp in the case of legends which must always
function.

This type of display, developed by the author at Rank Precision
Industries Ltd, is competitive with conventional designs since mass-

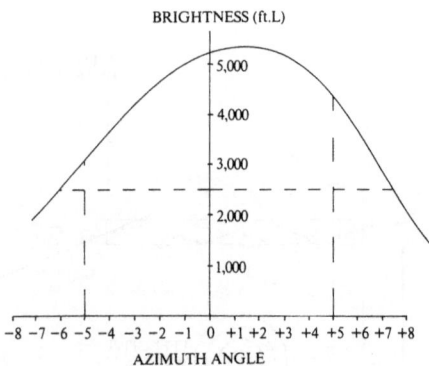

BRIGHTNESS (ft.L)

AZIMUTH ANGLE

Fig. 25 The variation of the brightness of an output aperture in the motorway sign as a
function of azimuthal angle. (The specification requires a brightness of at least 2500 ft L
at $\pm$ 5 deg.).

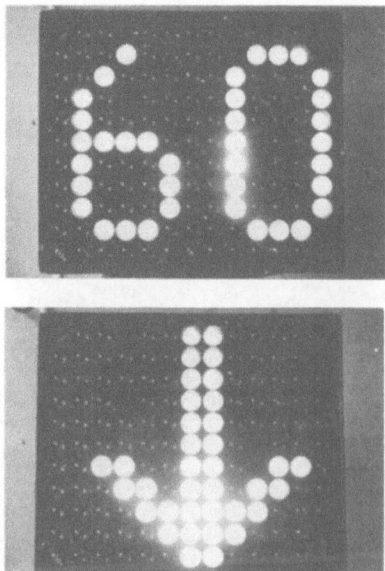

Fig. 26   Photograph of production motorway sign showing:
(a) a speed restriction, (b) a direction indication. (Courtesy of the Department of the Environment)

production techniques were developed for the manufacture of the branched light guides. This was possible owing to semi-continuous production of fibre bundles, as described in section 4.II.A. A typical branched guide for incorporation into this display is shown in Fig. 27, where the bundles are made in three lengths to reduce the weight of the finished sign. Up to sixteen such guides can be used in one sign which involves a total bundle length of around 450 m.

C. High-Contrast Displays

The above approach is ideally suited to any sign operating on the matrix principle. However, some signs are not amenable to this approach and, in certain instances, the matrix display is not aesthetically acceptable. The usual approach in these cases is to use a mask technique, illustrated in Fig. 28. In this, the light from the lamp is collimated by a parabolic mirror, whose aperture exceeds the maximum dimension of the legend. The light passes through a mask which defines the legend and is diverged by means of a lenticular screen into the required angular spread. This screen can be coloured to produce a coloured legend. Obviously, only one legend can be displayed by this means.

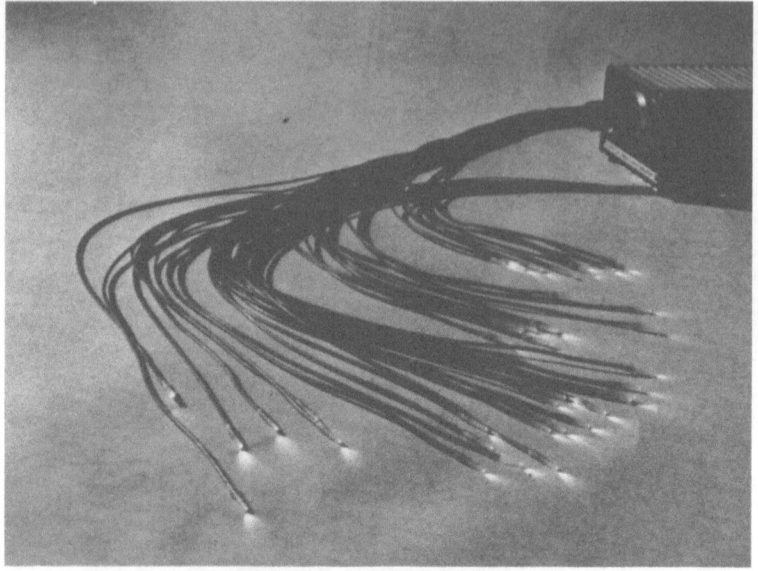

Fig. 27 Photograph of a branched light guide used to take the light from a single lamp to each matrix point, required in any given legend, in the motorway warning sign. (Courtesy of Rank Precision Industries Ltd)

There are two main disadvantages of such a system in conditions of high ambient lighting. Firstly, the reflection of ambient light from the screen reduces the visibility of the legend. Secondly, ambient light can pass through the screen and be reflected back through the system to a viewer who will see the legend even when the lamp is switched off;

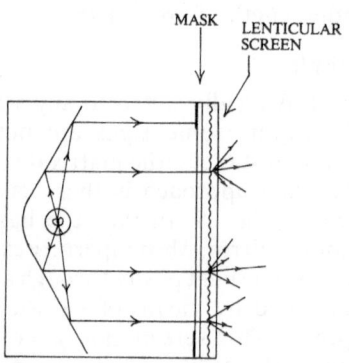

Fig. 28 Typical optical system used in hazard warnings signs for high angle viewing.

this effect is known as a 'phantom' image. Both of these disadvantages are overcome if the light is spread over the mask area using a multi-tailed fibre bundle. To give the appearance of uniformity of illumination at close distances the outputs of the bundles are made small and pitched closely together. In a typical bundle, the outputs would be 100 μm in diameter pitched at 1.0 mm, set in blocks of a black plastic. Since the fibres occupy only 1% of the total mask area the 'phantom' effect must be reduced by at least a factor of 100, since 99% of the light striking the surface will not enter the system. In fact, the reduction is much greater than this, since the N.A. of the fibres is about 0.5 so that only 25% of the ambient light entering the system will be accepted by the fibres.

In the manufacture of this type of display, bundles of fibres are laid, with one end fixed at 1.0 mm pitch, on to a plastic block, whose length equals that of the required legend. These ends are at right angles to the length of the block. The block width is 2.0 cm and its thickness is 0.9 mm, so that correct pitching at right angles to the block is automatically achieved when blocks are stacked together (Fig. 29). The number and position of the fibres in each block is varied so that no masking is required, the desired legend being created by the fibre distribution within the assembled sign. The finished blocks are stacked, bonded, and the output face, which is to be the output face of the completed sign, is fine-ground to reduce the amount of specular reflection. The free ends of the fibre bundles are compacted into a ferrule and finished as in the case of a normal light guide.

The completed sign provides legends of high light output over wide angles, up to 70° to the normal, and has found uses in the display of hazard warnings to motorists and pedestrians in urban areas, where a high viewing angle is necessary. A typical example of a sign of this type is shown in Fig. 30, and is used on pedestrian crossings. In this

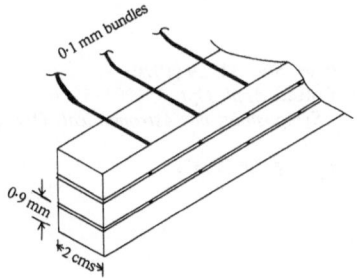

Fig. 29   Layout of fibre bundles in building up a high contrast, high angle hazard warning sign.

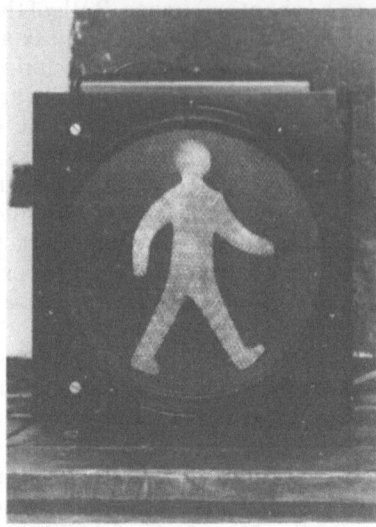

Fig. 30 Photograph of a hazard warning sign for pedestrian crossings, using fibre optics. The absence of 'phantom' images should be noted. (Courtesy of Rank Precision Industries, Ltd)

particular design a coloured legend is required and the front face of the fibre block has been covered by a lenticular screen of coloured plastic. Since the bundles occupy only 1.0% of the area of the output, it is obvious that a number of other legends can be interlaced into this type of sign without bundle overlap. Such signs have been proposed with two alternative direction indicators to alter traffic flow during peak periods without the normal risk of 'phantom' images appearing in conditions of bright sunlight.

REFERENCES

1. J. W. T. Walsh, *Proc. Phys. Soc.* **32**, 59 (1920)
2. W. L Stahl and R. H. Potter, *Appl. Opt.* **5**, 1203 (1966)
3. N. S. Kapany, *Proc. Symposium on Astronomical Optics*, Manchester (1955), p. 288
4. M. J. Forrest, *J. Sci. Instr.* **44**, 26 (1967)

CHAPTER 6

Coherent Bundles – Basic Theory

In the previous chapters the properties of systems have been discussed in which the optical fibre assembly has been used purely as a means of transferring light flux between a number of points. In these systems, little heed was paid to the optical isolation of a fibre from its neighbours, i.e. the fact that the light output from an optical fibre is a function solely of the light input to that fibre. Because of this isolation, any variation in the intensity of light across the input end of a light guide will manifest itself as a variation in intensity across the output face. However, these variations will not correlate simply, since the position of any fibre in the input face of the bundle is not related in a simple manner to its position in the output face.

In coherent bundles, on the other hand, the fibres are assembled so that the relative positions of the ends of any fibre within the bundle are simply related. These bundles are used when information is required about the spatial variations of intensity of the light distribution on the input face. Perhaps the best known example of such a bundle is the image-transferring bundle in which the fibre distributions at the ends are mirror images of each other. An image formed on the input face of a bundle of this type will be seen on the output face, although some degradation will take place during the transfer (Fig. 1). In this chapter we will consider the theory underlying this use of fibre optics and discuss the effect this has on the design of coherent bundles. It will be useful to confine our remarks to the image-transferring bundle although this in no way alters the general validity of the theory.

I. SPATIAL RESOLUTION OF AN OPTICAL FIBRE

An optical fibre will collect all the light within its N.A. falling on the end face of its core. If the length of this fibre is very much larger than the core diameter, as is invariably the case, then the distribution of light across the output face will be uniform, even if the distribution over the input face were non-uniform. This is due to the many reflections experienced by the light beam and the integrating effect of the skew rays, as was discussed in Chapter 2. A fibre cannot therefore be expected to transfer spatial information of detail finer than its core

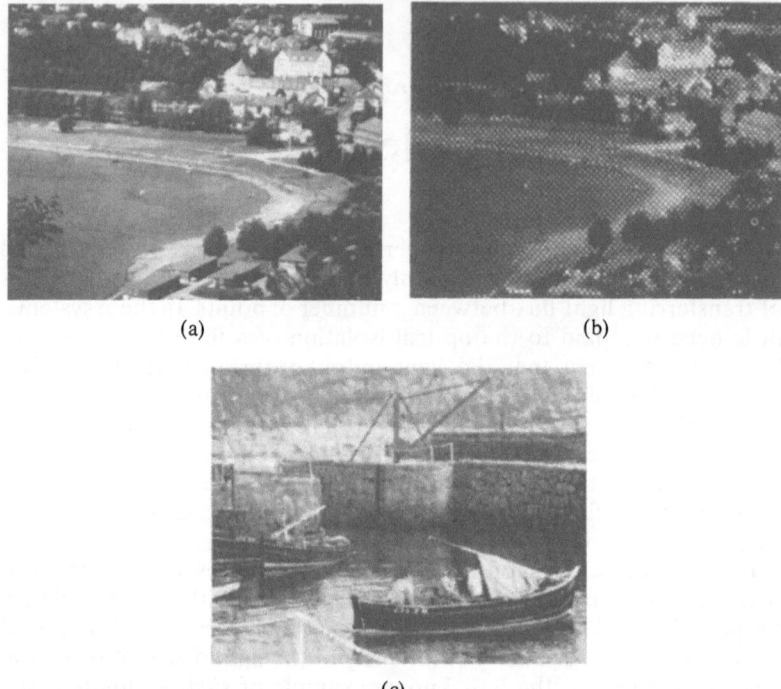

(a)   (b)

(c)

Fig. 1   Image seen through a coherent fibre bundle showing the degradation in the image due to the transfer.
(a) Original image.
(b) Image, in (a), seen through a coherent bundle with 70 × 50 fibres. (Courtesy of Barr and Stroud Ltd)
(c) Image transferred through a coherent bundle with 170 × 130 fibres. (Courtesy of Barr and Stroud Ltd)

diameter and can be regarded as being a single information point within the bundle.[1] In a coherent bundle these points are spaced one fibre diameter apart and information theory predicts that in such a case the maximum spatial frequency which can be transferred is $1/2d$, where d is the fibre diameter. This is illustrated for the case of image transfer in Fig. 2.

To consider in more detail the transfer of spatial information, we will discuss the transfer of a line pattern which varies sinusoidally in intensity with distance along the x-axis. The reason for choosing this particular distribution is that any intensity distribution can be analysed, by Fourier analysis, into a series of sinusoidal variations of differing frequencies. Such a distribution is shown in Fig. 3 and we will represent

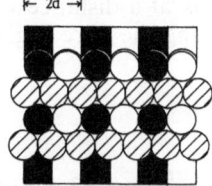

Fig. 2 Diagrammatic representation of the transfer of a bar chart through a coherent fibre bundle (fibre diameter, d) at the limiting spatial frequency, (shaded fibres are 50% illuminated).

it by:

$$I_1 = \frac{1}{2} I_0 \left(1 + \cos \left(2\pi \frac{x}{2t} \right) \right)$$

where I_1 is the intensity at a distance x from the origin, $2t$ is the periodic length and I_0 is the maximum intensity. This can be written as:

$$I_1 = \frac{1}{2} I_0 (1 + \cos (2\pi\omega x)) \tag{1}$$

where ω is the spatial frequency.

Let us consider the light output from a fibre whose input is positioned on the axis, at a distance x from the origin. For reasons of simplicity we will assume that the fibre has a square cross-section with one pair of sides parallel to the axis. If the core side measurement is d and we consider a rectangular element of width dx, at a distance x from the origin, then the amount of light entering through this element is given by:

$$dF = \frac{1}{2} I_0 (1 + \cos (2\pi\omega x)) \, dx \times d \tag{2}$$

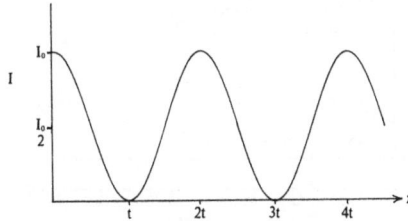

Fig. 3 Variation in intensity (I) of the line image used in the theoretical consideration of image transfer through a coherent bundle.

If the centre of the fibre is at a distance x from the origin then the total amount of light entering the fibre is given by:

$$F = \int_{x-\frac{d}{2}}^{x+\frac{d}{2}} dF$$

$$= I_0\left\{\frac{1}{2}d^2 + \frac{d}{2\pi\omega}\cos(2\pi\omega x)\sin(\pi\omega d)\right\} \tag{3}$$

Since this light is spread uniformly over the output face, the intensity of the output beam is given by:

$$I_2 = \frac{F}{d^2}$$

$$= I_0\left\{\frac{1}{2} + \frac{1}{2\pi\omega d}\cos(2\pi\omega x)\sin(\pi\omega d)\right\} \tag{4}$$

Figure 4 shows the relation given in Eq. (4) plotted for various values

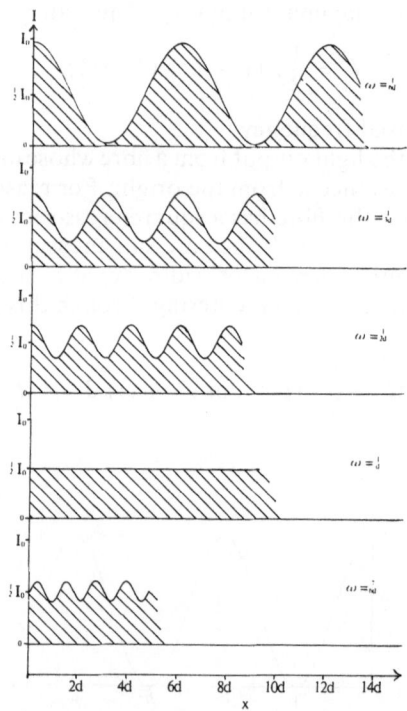

Fig. 4 Intensity of light (I) transmitted through a square single fibre (side d), for various spatial frequencies, plotted as a function of distance along the x-axis.

of spatial frequency. It will be noted that for $\omega = 1/d$ there is no variation in output intensity with fibre position, i.e. zero spatial resolution. It is interesting to note that there is some variation in intensity for spatial frequencies above this zero position. This variation is out of phase with the input light distribution and is analogous to the spurious resolution which occurs with lenses and is illustrated diagrammatically in Fig. 5.

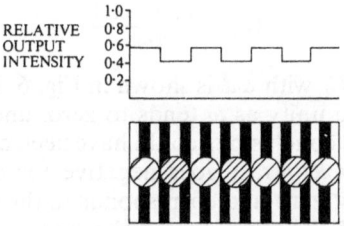

Fig. 5   Diagram illustrating spurious resolution through a coherent bundle for $\omega = \dfrac{3}{2d}$.

Neighbouring fibres have output intensities approximately in the ratio 1.5:1, with the highest intensity centred on a black bar. The relative intensity issuing from each fibre is also shown.

II. IMAGE TRANSFER THROUGH A COHERENT BUNDLE

If we now consider a single layer of parallel square fibres which are touching and extend along the x-axis, then the output from each fibre will be obtained by substituting for its centre distance in Eq. (4). A useful measure of the effectiveness of this system is obtained by making use of Michelson's visibility criterion given by:

$$C = \frac{I_{max} - I_{min}}{I_{max} + I_{min}} \qquad (5)$$

where I_{max} is the maximum intensity in the output and I_{min} is the minimum intensity in the output.

This criterion was used by Michelson to determine the visibility of interference fringes but is now in general use in optics and is called the contrast of the image. It will be seen that the contrast has a maximum value of unity, when I_{min} is zero, which is the optimum visibility condition, and a minimum value of zero when $I_{max} = I_{min}$ in which case the visibility is zero.

It is obvious from Eq. (4) that, if we assume zero sheath thickness, the required values are:

$$I_{max} = \frac{1}{2} + \frac{\sin(\pi\omega d)}{2\pi\omega d}$$

and

$$I_{min} = \frac{1}{2} - \frac{\sin(\pi\omega d)}{2\pi\omega d}$$

If we substitute these values in Eq. (5) we obtain

$$C_1 = \frac{\sin(\pi\omega d)}{\pi\omega d} \qquad (6)$$

The variation of C_1 with ωd is shown in Fig. 6. It will be seen from this that C_1 tends to unity as ω tends to zero, and that C_1 is zero at $\omega = 1/d$; both these conclusions could have been drawn from an intuitive standpoint. The contrast goes negative for certain values of ω greater than the first zero and corresponds to the spurious resolution illustrated in Fig. 5. The reason for the negative value is that the positions of maximum and minimum are interchanged at the outputs of the fibre, which is seen clearly in Fig. 5.

However, the values of contrast predicted by Eq. (6) can only be obtained with the fibre assembly if there are fibres whose centres correspond to the stationary positions of the pattern. This will not, in general, be true and any other configuration will give lower values

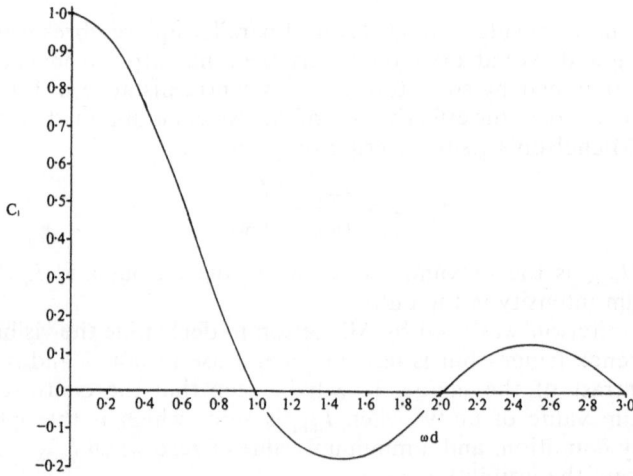

Fig. 6   Variation of the contrast function C_1 plotted against spatial frequency (ωd).

for this contrast, the worst case being when the centres of the appropriate fibres are displaced $\frac{1}{2} d$ from the stationary points. In this case, there will be two adjacent fibres on either side of each stationary point with equal outputs. Therefore the intensity of light issuing from these will be the same as would be obtained from an array of fibres, side $2 d$, having a fibre centrally disposed at each stationary point. In this case, the contrast is given by:

$$C_2 = \frac{\sin 2\pi\omega d}{2\pi\omega d} \tag{7}$$

C_2 is zero at $\omega = \dfrac{1}{2 d}$, which is illustrated in Fig. 7.

It will be appreciated therefore that there is no unique value of contrast at a given spatial frequency for a fibre bundle; the actual value which is obtained will vary between that given by Eq. (6) and that given by equation (7). This is illustrated in Fig. 8 which shows contrast measurements taken on a single bundle for different positions of the bundle input relative to the test pattern. It will be seen from this that C_2 is the contrast function which must be used in any design consideration.

There is one set of circumstances under which unique contrast measurements can be made; this is the case where the bundle ends are moved in synchronism relative to the test pattern. Since the ends are in synchronism, the transmitted image is stationary and contrast

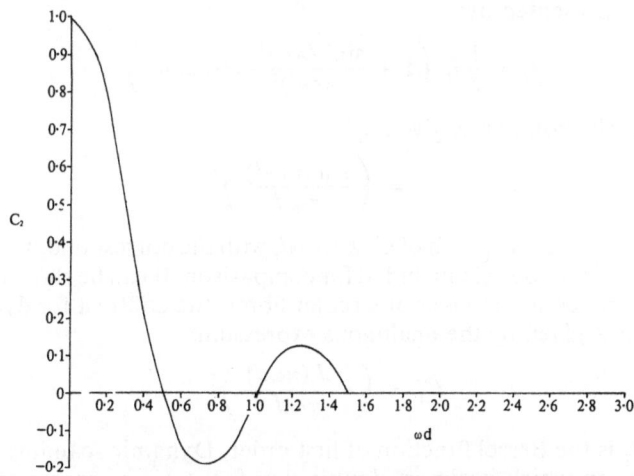

Fig. 7   Variation of the contrast function C_2 plotted against spatial frequency (ωd).

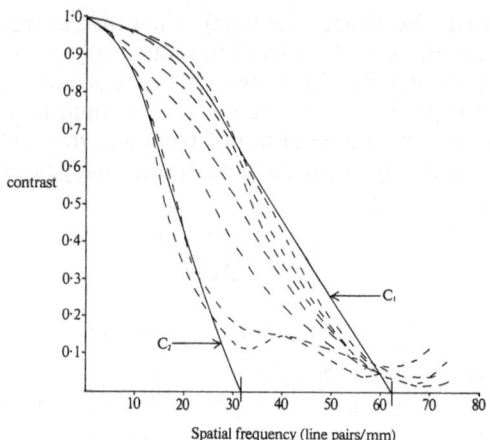

contrast

Spatial frequency (line pairs/mm)

Fig. 8 Measured variation in contrast taken at different positions on a coherent bundle with 16μ fibres. The theoretical contrast functions C_1 and C_2 are shown as solid curves.

measurements can be made. If the motion of the bundle is rapid compared to the response time of the detector, then the output at any point will be the time-averaged output of a fibre as it passes this point. If the motion is uniform, this is equivalent to averaging the output from one fibre from $x - d/2$ to $x + d/2$, where x represents the distance of the point of measurement from the origin. It is a simple matter to show that under these conditions the transferred image of the test pattern can be represented by:

$$I_3 = \frac{1}{2}I_0\left(1 + \frac{\sin^2(\pi\omega d)}{(\pi\omega d)^2}\cos 2\pi\omega x\right) \tag{8}$$

and that the contrast is given by:

$$C_3 = \left(\frac{\sin(\pi\omega d)}{\pi\omega d}\right)^2 \tag{9a}$$

Fig. 9 shows the variation of C_3 with ωd, with the corresponding curves for a stationary bundle included for comparison. It can be shown[2] that, for a single coherent layer of circular fibres, the contrast for dynamic scanning is given by the analogous expression:

$$C_3 = \left(\frac{2 J_1(\pi\omega d)}{\pi\omega d}\right)^2 \tag{9b}$$

where J_1 is the Bessel function of first order. Dynamic scanning yields a resolution which is almost double that for static measurements, in the worst case, i.e. using C_2 as the contrast function.

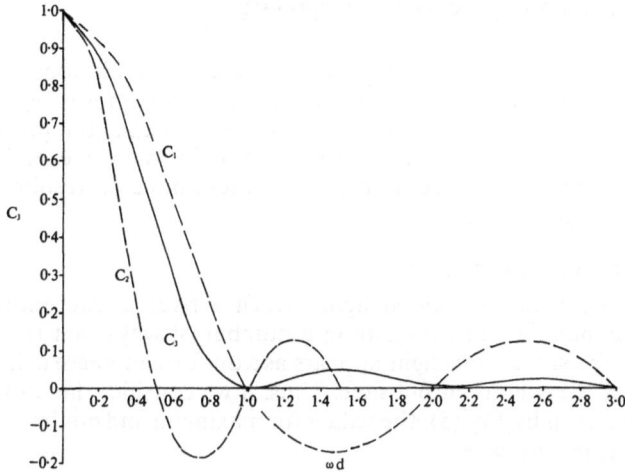

Fig. 9 Variation of the dynamic contrast function C_3, plotted as a function of spatial frequency (ωd). The static contrast functions, C_1 and C_2 are shown dotted for comparison.

In practical assemblies the sheath thickness is finite but is small compared to the core diameter, in which case the effect of sheath thickness is negligible for a moving bundle. With a stationary bundle one must use the total fibre diameter instead of the core diameter in the formulae for contrast.

The bundles which have been discussed act in one dimension only, viz along the x-axis. If, as is usually the case, a bundle is required which transfers a two-dimensional distribution of light, then this can be formed by stacking coherent layers of fibres. This in no way alters the theory which has been developed above. It can be shown[3] that the dynamic response of a coherent bundle made with square fibres is given in terms of the following contrast function:

$$C_4 = \left\{ \frac{\sin (\pi \omega d \cos \theta)}{\pi \omega d \cos \theta} \cdot \frac{\sin (\pi \omega d \sin \theta)}{\pi \omega d \sin \theta} \right\}^2 \qquad (10)$$

where θ is the angle which the sinusoidal distribution of light at the input makes with the sides of the fibres. For $\theta = 0$, Eq. (10) reduces to Eq. (9a), whilst for $\theta = \pi/4$ the contrast has a minimum value, which corresponds to the performance of a bundle with a fibre size of $\sqrt{2}\, d$, which could have been predicted qualitatively.

III. DEGRADATION OF CONTRAST

Practical coherent bundles never achieve the resolutions predicted by the theory outlined above, although the discrepancy is negligible in most instances. There are two main causes for this, the presence of stray light and structural defects in the bundle. We will examine both of these causes and discuss the effect these have on resolution and means of minimizing this.

A. Effects of Stray Light

Stray light, i.e. unwanted light, which arrives at the output of a coherent bundle can originate in a number of ways; but the overall effect is the same – this light appears as a uniform increase in illumination over the whole output face. Thus, if we consider the formula for contrast given by Eq. (5), the values for maximum and minimum intensities are, in this case:

$$I'_{max} = I_{max} + I_s$$

and

$$I'_{min} = I_{min} + I_s$$

where I_s is the stray light intensity. Substituting the above values in Eq. (5) we obtain:

$$C_s = \frac{I_{max} - I_{min}}{2 I_s + I_{max} + I_{min}} \qquad (11)$$

where C_s is the contrast under the influence of stray light. It is immediately obvious from Eq. (11) that the resulting contrast is less than in the case with zero stray light. If the contrast with zero stray light were 100% (i.e. $I_{min} = 0$) then C_s is given by:

$$C_s = \frac{I_{max}}{2 I_s + I_{max}}$$

$$= \left(1 + \frac{2 I_s}{I_{max}}\right)^{-1} \qquad (12)$$

Figure 10 shows the variation of C_s with the relative amount of stray light present.

One area in which the question of stray light is of utmost importance is where fibre optics is used to provide a face-plate in opto-electronic vacuum tubes, e.g. C.R.T.s. In this application the image is formed in a phosphor deposited directly on the end-faces of the fibres and behaves as a Lambertian source. In such a system stray light originates

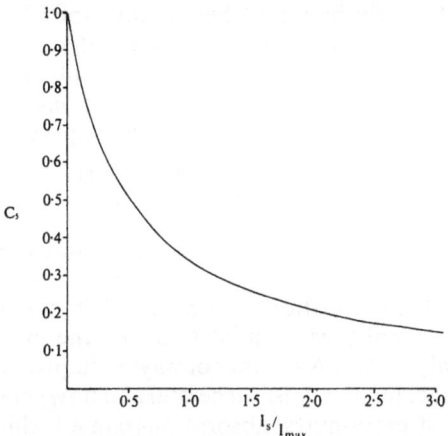

Fig. 10 Variation of the contrast (C_s) of an image with stray light present, plotted against the relative amount of stray light (I_s/I_{max}).

from two main sources. First, if the N.A. of the fibres is less than unity then some of the light entering the fibre cores is not accepted by the fibres and, secondly, light which enters the bundle via the sheath will not be accepted by any fibre. Light falling into either category will suffer multiple scattering at the core–sheath interfaces during its passage through the bundle and will tend to emerge as a uniform illumination.

In a typical bundle, some 10% of the face is occupied by sheath material, so that 10% of the incident light will appear as stray light. In addition, it will be recalled from Chapter 2 that the fraction of light from a Lambertian source which is accepted by a fibre is given by (N.A.)². Thus the fraction of stray light is given by $1 - (N.A.)^2$. For a bar-chart image, the intensity of stray light is given by:

$$I_s = \frac{1}{2} \left\{ 0.1 + 0.9 \left(1 - (N.A.)^2 \right) \right\}$$

and the maximum intensity (neglecting I_s) is given by:

$$I_{max} = 0.9(N.A.)^2$$

The contrast of this system is therefore obtained from Eq. (12) and is given by:

$$C_s = \frac{0.9(N.A.)^2}{0.9(N.A.)^2 + 0.1 + 0.9(1 - (N.A.)^2)}$$
$$= 0.9(N.A.)^2 \qquad (13)$$

This contrast, it will be remembered, corresponds to a contrast of unity without stray light. It can be shown that the corresponding formula for other conditions is given by multiplying Eq. (13) by the diameter-dependent expression derived earlier, e.g. Eq. (6). We thus obtain for the above system the final contrast expression:

$$C_s = 0.9(\text{N.A.})^2 \frac{\sin(\pi\omega d)}{\pi\omega d} \qquad (14)$$

Figure 11 shows the variation of C_s spatial frequency for a range of values of N.A.

In this particular application it is obvious that stray light is a serious problem. For example, with an N.A. of 0.5 the maximum contrast achievable is only 22.5%. A number of ways of improving this situation have been devised, but the most successful and now generally-accepted method is that of extra-mural absorption (e.m.a.) which, as the name implies, is based on absorbing the stray light within the bundle. Obviously, any absorbing material which is introduced into the bundle must not affect the useful light passing through the bundle and it is normally placed outside the sheath. A convenient way of doing this is to place a second sheath, which is absorbing, around each fibre. In such a case the path of a typical ray is shown in Fig. 12 and it can be shown that the path length in the e.m.a. is given by (it has been assumed that the refractive index of the e.m.a. sheath equals that of the clear sheath):

$$P = \frac{n_2\, t\, L}{\cos\phi_2(T \tan\phi_1 + (a - T)\tan\phi_2)} \qquad (15)$$

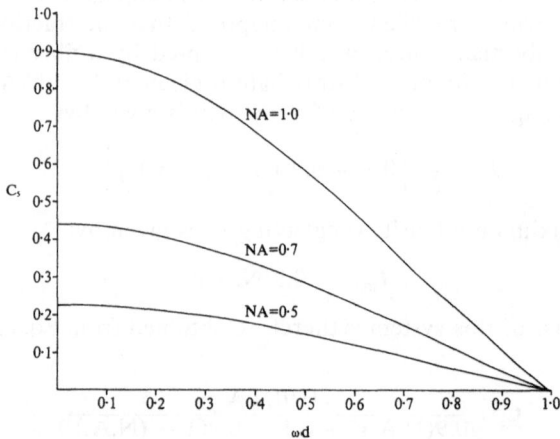

Fig. 11 Variation in contrast (C_s) for a Lambertian image plotted against spatial frequency ωd for various N.A.'s.

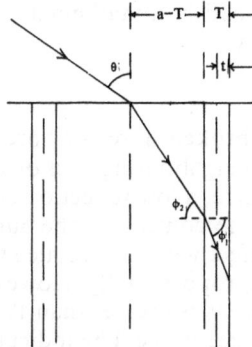

Fig. 12 Diagram of the path of a light ray through an optical fibre where the ray is not accepted by the fibre.

where: t is the e.m.a. sheath thickness,

T is the combined thickness of the e.m.a. sheath and the clear sheath.

L is the bundle length,

$2a$ is the total fibre diameter,

and ϕ_1 and ϕ_2 are the refracted angles in the e.m.a. sheath and fibre respectively.

In practice, ϕ_1 and ϕ_2 are almost equal so that we can write with sufficient accuracy:

$$P = \frac{n_2\, t\, L}{a \sin \phi_1} = \frac{n_2\, t\, L}{a \left(1 - \dfrac{\sin^2 \theta}{n_1^2}\right)^{\frac{1}{2}}} \tag{16}$$

Thus we can arrange any degree of absorption of stray light by a suitable choice of e.m.a. sheath thickness. It should be noted that this method applies equally to the two sources of stray light mentioned earlier.

However, for various reasons which will be discussed later, it is convenient to use doped glass as an e.m.a. material, and this means that the bulk absorption of the material is not as high as one would like. If a high degree of absorption is required, the e.m.a. sheath thickness becomes inordinately large, and the core packing fraction drops considerably. In practice a compromise is reached which varies according to the particular requirements; this will be discussed in detail in a later chapter.

B. Effects of Structural Defects

Practical coherent bundles suffer from a number of structural defects which degrade the optical performance. The effect of these cannot readily be described in a mathematical fashion, and will only be discussed qualitatively.

Opaque Areas

These are caused by broken fibres or fibres whose transmission is significantly lower than neighbouring ones owing to an imperfect core-sheath interface and, hence, a low reflection coefficient. The presence of these will not affect the contrast of the bundle if their distribution is random, since their effect will be to reduce the maximum and minimum values of intersity proportionally. However, it is found that if the proportion of opaque areas is greater than 1% then the quality of the transferred image is not acceptable. This indicates that perhaps contrast measurement is of less value in fibre optics than in lens optics.

Displaced Fibres

It is found that in most coherent bundles there are fibres which have been displaced from their ideal position during the manufacturing process. The effect of this will be to reduce the contrast of the bundle since the displaced fibres can transfer light from a region of maximum intensity to a region of minimum intensity. In general, the amount of displacement is only a few fibre diameters so that the effect of this will only be noticeable in the higher spatial frequencies, where the contrast will decrease more rapidly than the theory predicts and will become zero at a frequency lower than the theoretical value. If the fibres are uniformly distributed within a distance of δ from their ideal positions then the image of a square wave will be transferred as a trapezoidal wave with a base width $D + 2\delta$, where D is the width of the square wave. The effect of this as the frequency of the square wave is increased is shown in Fig. 13, where it will be noted that the transmitted contrast

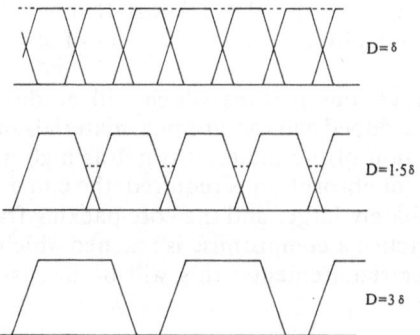

Fig. 13 Transmitted image of square wave test chart through a coherent bundle with imperfect fibre positioning. This is shown for three values of square wave width (D). Dotted line shows resulting light distribution.

is zero when $D = \delta$, and is unaffected by this defect when $D > 2\delta$. For intermediate frequencies, the theoretical contrast must be multiplied by a factor $(D - \delta)/(3\delta - D)$; the effect of this on the contrast is shown in Fig. 14 for δ equal to one, two and three fibre diameters. In some instances the displacement of the fibres is not random but has a periodic nature which can have a more serious effect on the final contrast at frequencies close to that of the displacement. The cause and effect of these variations from the ideal situation will be outlined in the next chapter.

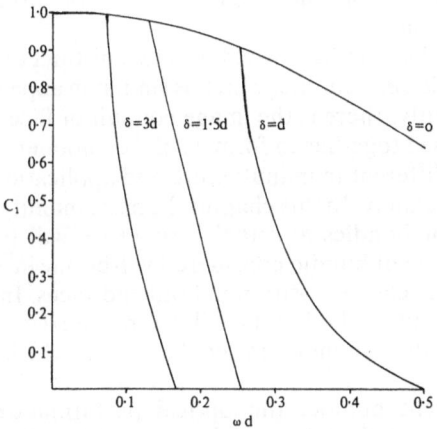

Fig. 14 Variation contrast (C_1) with spatial frequency (ωd) for coherent bundles with the fibres randomly distributed within a distance of δ from their true positions. This is plotted for $\delta = 0$, d, $2d$ and $3d$, where d is the fibre diameter.

It will be appreciated from the above that the measurement of contrast does not define completely the performance of a coherent bundle and only gives a unique figure for the case of dynamic scanning. For this reason contrast measurements have to be supplemented with a description of the type and quantity of bundle defects before the final optical performance can be assessed. In fact, in most instances, a satisfactory assessment of performance can be made with only a knowledge of the fibre diameter plus the defect distribution.

REFERENCES

1. N. S. Kapany and J. J. Burke, *J. Opt. Soc. Am.* **52**, 1351 (1962)
2. P. G. Roetling and W. P. Ganley, *J. Opt. Soc. Am.* **52**, 99 (1962)
3. N. S. Kapany, *Fiber Optics*, Academic Press Inc., New York, (1967), p. 363

CHAPTER 7

Coherent Bundles – Manufacture and Properties

In the following two chapters, the properties and applications will be described of optical fibre bundles that employ the principles outlined in the previous chapter.

Coherent bundles can be divided into two distinct types. First, there is the flexible coherent bundle, which is similar in appearance to a light guide and secondly, there is the image conduit or face-plate, in which the fibres are fused together to form a solid component. These types of bundles are so different in manufacture and application that they will be discussed seperately. In this chapter the basic manufacturing techniques of coherent bundles and their properties will be outlined; for simplicity, the type of bundle considered will be one in which the fibres occupy the same relative position in both end-faces. In this, an image formed on one end of the bundle will be transferred to the other end with a definition dependent upon the diameter of the individual fibres in the bundle.

In any coherent bundle, the optical performance is controlled basically by the fibre diameter and, because of this, diameters are used which are much smaller than would be demanded by the criterion of flexibility. Fibre diameters in these bundles can be as low as 5 μm and this would create almost insuperable handling problems if such fibres had to be assembled individually. However, techniques have been developed for the manufacture of combinations of such fibres which can be handled as a single unit, but which still behave optically as separate fibres; these are generally called multiple fibres.

I. MULTIPLE FIBRES

A multiple fibre consists of a group of individual optical fibres which are fused together by means of their sheaths to form a single mechanical unit, but without affecting the optical performance and isolation of the individual fibres within the group. These multiple fibres are used as the basic units in the manufacture of large bundles and should therefore have a cross-section which permits efficient packing, so that any voids are much smaller than the individual fibre diameter. Since the fibre diameter is much smaller than the overall diameter of the multiple fibre, circular cross-sections are not used as these would produce voids of about one-third the total multiple fibre diameter in size. Since the individual fibres can be as small as one four hundredth of the multiple fibre diameter such a situation is clearly unacceptable. Generally, square or hexagonal multiple fibres are used, since these cross-sections should not produce voids at all when suitably packed together.

These multiple fibres are manufactured using a drawing technique similar to that employed in the manufacture of single fibres (Section 3.1). A number of single fibres of relatively large cross-section (1 mm to 3 mm in diameter) are stacked into a jig to yield a bundle of the desired cross-section, but much larger than the required size of the multiple fibre. The fibres are aligned during the stacking to be parallel to one another and the resulting stack might be typically 50 mm across. Since the fibres in the stack are relatively rigid units, the stacking and aligning of these is straightforward. This stack is then mounted in a drawing unit similar to that used for single fibres, and the cross-section reduced by a drawing process to the required size, during which process the single fibres are bonded together into a single physical mass by the fusion of their sheaths.

In this drawing process, the drawing temperature and the temperature gradient within the furnace are much more critical than in the single fibre case, for the following reasons. First, the furnace temperature must be high enough to allow drawing, but must also be low enough to ensure that the viscous forces prevent any significant change in cross-section. This change would be caused by the action of surface-tension forces on a non-circular cross-section which would act to render it circular. Secondly, the furnace temperature and temperature gradient must be such that the fusion of the sheaths is uniform throughout the stack and, if required, the glass flows to fill the interstitial voids forming a completely solid unit. Thus the furnace design and temperature control are much more critical than in a single-fibre process, but the correct parameters can be set up, and maintained, to produce multiple fibres whose cross-section is virtually that of the initial stack and in which uniform fusion of the individual fibres has been achieved (Fig. 1). It should be noted that, since neighbouring sheaths are fused together, the sheath thickness on the single fibre can be half the 'penetration depth' required to give optical isolation.

The above technique works very well with stacks in which the fibres are hexagonally packed. However, in certain applications square multiple fibres with square packing are required, in which case the interstitial voids are larger and it is difficult to prevent a change in cross-section, a 'barrel-shaped' cross-section being the usual result. This can be prevented by the introduction of small fibres into the

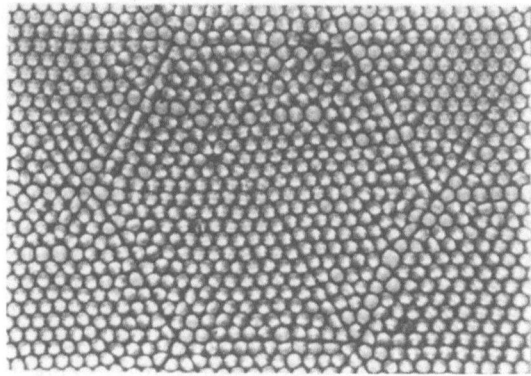

Fig. 1 Photomicrograph of a hexagonal multiple fibre within a fused plate. The 'star' defect caused by a missing fibre can be seen close to centre of the bundle. The individual fibre size is 50 μ.

interstices of the initial stack thus greatly reducing the volume occupied by voids. These interstitial fibres are about one-third of the diameter of the main fibres (Fig. 2).

This drawing process can be repeated with stacks of multiple fibres to produce 'double' multiple fibres, in which very fine fibres can be produced in relatively large mechanical units (say 5 μm fibres in a stack 2 mm across).

Fig. 2 Photopraph of a square multiple fibre, with interstitial fibres to maintain cross-section, main fibre size 75 μ. (Courtesy of Barr and Stroud Ltd)

II. THE INCORPORATION OF EXTRA-MURAL ABSORPTION

As was discussed in the previous chapter, there are some applications for coherent bundles in which the N.A. of the input is higher than that of the fibres. Under these conditions, the light which is not accepted by the fibres can cause a serious reduction in the contrast of the final image, but this effect can be reduced to negligible proportions by the use of absorbing material surrounding each fibre. This technique is known as extra-mural absorption (e.m.a.) and is incorporated into the multiple fibre in one of two ways.

The first of these is the obvious one of surrounding each fibre by a second sheath of absorbing glass. As was discussed in the previous chapter, the thickness of the second sheath is determined by the aspect ratio required of the fibre in use and by the volume absorptivity of the material comprising this second sheath. Obviously, if the light within the N.A. of the fibre is not to be affected by this second sheath, then the first sheath, i.e. the transparent one, must be thicker than the 'penetration depth' of the light into this sheath during reflection. This requirement doubles the amount of clear sheath required in the multiple fibre, since the sheath around each fibre must now be equal to the penetration depth, with a consequent reduction of core packing fraction. This is further reduced by the fact that the absorbing glass has a finite thickness, which is normally similar to that of the clear sheath.

The thickness of e.m.a. is determined by the amount of absorption required in the e.m.a. glass, which has a low volume absorptivity compared to those found in metals. This is because there is a limit to the amount of colouring ions that can be introduced into a glass, determined by the level at which the ion starts to cause devitrification of the resulting glass. With the colouring ions that are normally used, the resulting glass has an absorptivity which requires a relatively thick second sheath for effective operation of the e.m.a.

Since there is a lower limit to the permissible thickness of clear sheath, the core packing fraction falls dramatically as the core size is reduced. Figure 3 shows the variation of core packing fraction with fibre diameter, assuming a clear sheath thickness of 0.5 μm and an e.m.a. thickness of 0.3 μm. It will be seen from this graph that a fibre with a 5 μm diameter has a core packing fraction of 46% as opposed to 64% for the same fibre with no e.m.a. It is essential to use the overall diameter for comparison purposes, since it is this which determines the number of fibres per unit area, i.e. the information capacity of the assembly. Because of this, the above method of applying e.m.a. is normally only used with relatively large fibre diameters (say above 20 μm) where the increased area of sheath and e.m.a. material has a small effect on the transmission.

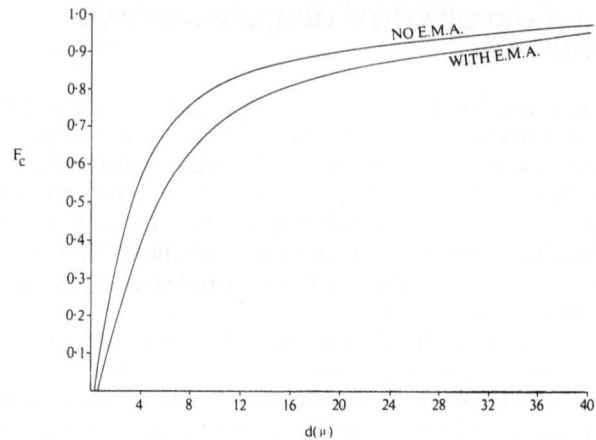

Fig. 3   Variation of core packing fraction (F_c) plotted as a function of fibre diameter (*d*) with a clear sheath thickness of 0.5 μ. The curve for an e.m.a. fibre includes an e.m.a. thickness of 0.3 μ.

The second method of applying e.m.a. has been developed for use with small fibres (down to 5 μm in diameter), and is known as the interstitial technique. In this, e.m.a. is introduced into the stack in the form of rods of absorbing glass which are placed in the interstitial voids between the fibres. The actual volume of e.m.a. glass introduced equals that in the previous method and this technique relies for its success on the fact that a ray which is not accepted by the fibre has to traverse many fibres before leaving the bundle, and will, on average, pass through the same thickness of e.m.a. glass as in the previous method. As a result of this construction, the fibre is not completely surrounded by e.m.a. glass; in fact, only a small fraction of the perimeter of the fibre is in contact with the e.m.a. glass (see Fig. 4), so that the clear sheath can be much thinner than in the previous case. In practice, the clear sheath thickness can be made equal to that required for fibres without e.m.a. with no appreciable loss in efficiency, and a 5 μm diameter fibre can be incorporated into a multiple fibre with 0.25 μm clear sheath and e.m.a., equivalent to a thickness of 0.5 μm with a core packing fraction of 60%. An additional advantage is that the e.m.a. glass can be formed into rods with a much higher colouring-ion content than would be possible if tubing were made, so it is possible to reduce the volume of e.m.a. glass in the multiple fibre whilst maintaining the same degree of absorption.

Since this method depends on the escaping ray traversing a large number of fibres before leaving the bundle, it is reasonable to expect that this becomes less valid, for a given length of fibre, as the fibre

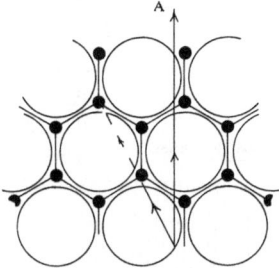

Fig. 4 Diagram showing the position of interstitial e.m.a. rods (black circles) in a multiple fibre and the projection of two typical rays through such a fibre.

diameter increases. It will be obvious that ray 'A' in Fig. 4 will not intercept any e.m.a. material if the fibre packing is perfect. Further, for a given length of fibre, the amount of light which passes through the plate in a similar fashion to ray 'A' will increase as the fibre diameter increases, since the chance of intercepting an e.m.a. rod will decrease. The use of interstitial e.m.a. is therefore restricted to the lower end of the range of fibre diameters used (below about 20 μm). It will be seen that these two techniques are complementary.

As in the case of single fibres, these multiple fibres can be produced in a wide range of cross-sectional sizes, from around 10 mm down to 50 μm or less.

III. MANUFACTURE OF FLEXIBLE COHERENT BUNDLES

A flexible coherent bundle is contructed from fibres, whether single or multiple, whose cross-section is small enough to render these flexible. The mode of construction is such that only the extremities of the fibres in the bundle are bonded together, so that, as in the case of light guides, the finished bundle is flexible. Typically, the cross-section of the fibres in such bundles ranges from 10 μm to 70 μm. Current methods of manufacture are variations of one basic technique, which will now be described.

A flexible fibre is drawn from a muffle furnace in the normal fashion and is wound on to a drum. The fibre passes through a system of guides, during its journey between furnace and drum, which accurately define the position of the fibre in space. The drum is traversed along its axis by one fibre diameter for each rotation which means that the fibre is collected on this drum as a close-wound helical layer (Fig. 5). If this process is stopped after a number of rotations and the layer removed

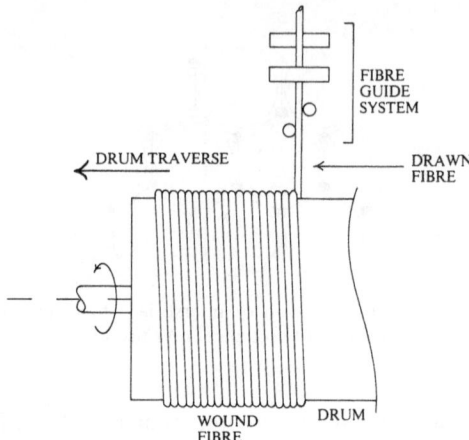

Fig. 5 Diagram illustrating the basic principle of winding a coherent layer of fibres, using a traversing drum.

by slitting the helix along a line parallel to the drum axis, an array of fibres is obtained in which the fibres are parallel to one another (see Fig. 6).

The method described above is basically the same as that used by Kapany[1] in the 1950s and was in general use, probably up to the mid-1960s. At about this time, the final assembly of coherent layers had been improved to such an extent that the quality of the finished product was being limited by defects in these layers. These defects stemmed from the impossibility of exactly matching the pitch of the traverse to the fibre diameter and, since a typical layer would contain about 150 fibres, any discrepancy in this matching would be magnified enormously. It was obviously unacceptable to have the traverse pitch less than the fibre diameter since this would eventually cause the fibre to lie on top of the preceding one. In practice, the traverse pitch was set to be slightly greater than the maximum fibre diameter, allowing for typical varia-

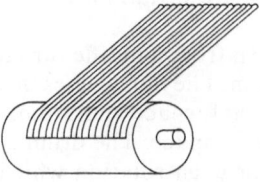

Fig. 6 A consolidated coherent layer being removed from the drum, after winding is complete.

tions in diameter during the run, and the fibres would be laid at a pitch of about 1.1 times their diameter. This was not a satisfactory arrangement and was unacceptable when multiple fibres were wound where the individual fibre diameter could be as small as one-sixth the overall size.

One solution to this problem will now be described, which was developed to a commercial proposition by the author at Rank Precision Industries Ltd.

. The underlying principle of this technique is to use, in effect, the fibre itself to set the traverse pitch. The basic features of the system are shown in Fig. 7, and consist of a stylus which runs on the drum surface and is tensioned to pull the fibre, as it comes on to the drum, across the drum surface to butt against the preceding fibre. The stylus is set slightly below the drum axis and the fibre first meets the drum at a distance of 1 to 3 mm from the layer. Because of this relatively large separation, the drum traverse need not be set very accurately and, indeed, for layers with widths less than 3 mm, the drum need not be traversed at all. The run is started by winding a few turns without the stylus to form a 'step' with a height of one fibre diameter and then the stylus is placed in position. The frictional forces between the fibre and the drum are fairly large, so this 'step' is not moved by the stylus when this is introduced. This technique produces high-quality coherent layers at a higher winding speed than is possible with the older technique, and layers have been made up to 30 cm wide using 40 μm fibre (i.e. 7500 turns of the drum).

This constitutes what may be termed a one-dimensional coherent

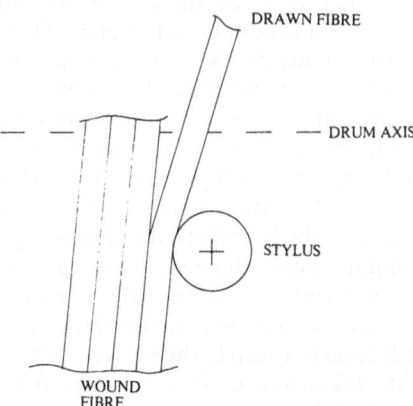

Fig. 7 Diagram illustrating the basic principle of winding a coherent layer of fibres, using a tensioned stylus.

bundle, i.e. this array will transmit the intensity variations of a line image coincident with one of the end-faces, within the resolution capability of the fibres. Obviously, if a number of these arrays could be stacked on top of one another, so that the fibres in each layer were accurately registered with respect to the fibres in adjacent layers at each end, then a bundle would be built up which would transmit a two-dimensional image, again within the resolution of the fibres.

There are two distinct methods of achieving this result, and these will now be discussed.

A. Coil-Winding Technique

In this technique, a single layer is wound on to the drum, as described above, until the layer width equals that of the desired bundle, and then the drum traverse is reversed whilst still winding the fibre. A second close-wound helix is thus built up on top of the first and the traverse is again reversed to start a third layer. This process is continued and the required bundle built up in a zig-zag fashion until the desired size of bundle has been produced. The fibres are then consolidated at one point, and the bundle removed from the drum by slitting through this point. In practice, the fibres are wound into a groove in the drum, this groove having the same width as the required bundle so that the side-walls of the groove support the layers of fibre during the winding process.

There are a number of practical difficulties associated with this technique. First, it is impossible to wind the layer to the exact width of the groove, and the later layers are therefore laid on a surface which is not level. This surface curves downwards at the centre if the actual layer width is larger than the groove width, since a build-up of fibres occurs in the groove at the exremities of each traverse. If the layer width is less than that of the groove a surface is formed which curves upwards at the centre. In either case, the coherence of the fibres at the edges is poor and the curved base tends to degrade the coherence of the later layers. Secondly, consolidation is difficult to achieve since the fibres are closely packed and any bonding agent must be introduced through the side of the bundle. An alternative method is to fuse the fibres together by heating a portion of the bundle to its softening point and applying pressure to this region. This can be successful although it creates two weak portions in the bundle, on either side of the fused region, where a rapid change in heat transfer rate occurs at the point where the fibres start to fuse, which creates strain in the glass during the annealing cycle after the fusion has taken place. Thirdly, the outer layers are longer than the inner layers, which creates problems in the final assembly of such a bundle into an instrument.

However, the most serious criticism of this technique is that the method of winding significantly reduces the resolution capabilities of the finished bundle. To understand the reason for this it is necessary to consider the winding of a single layer. As stated previously, this is a close-wound helix and when this is slit and taken off the drum a layer of parallel fibres is obtained. But these fibres are not wound normal to the drum axis owing to the pitch of the helix, and are in fact skewed a fibre diameter to the left or right, depending on the direction of drum traverse (see Fig. 5). If the development of the layer on to a plane were accurate, the image of a line focused on one end would be displaced by one fibre diameter at the other. This is obviously not important when we are considering a single layer, but in the technique just described adjacent layers have their direction of pitch reversed, so that two fibres in adjacent layers which touch at one end are spaced two fibre diameters apart at the other end, as depicted in Fig. 8. It will be seen from Fig. 8 that a test pattern has to have line widths of three fibre diameters to be resolved, in the best case, which means that the resolution of the bundle is reduced by a factor of three, in the direction of laying (Fig. 9).

Because of this, the above technique is no longer used, although a modification is still employed where the winding always takes place in the same direction. This is achieved by stopping the run at the end of each traverse, breaking the fibre, bringing the fibre and drum back to their original positions and re-starting the run. The practical difficulties still apply and others are introduced by the discontinuous nature of the modified technique. This renders the technique no more convenient than the following one.

B. Layer Stacking

In this technique, single layers are wound as described earlier, but are consolidated as single layers and removed from the drum. The coherent bundle is then build up by stacking these layers into a jig, which ensures that the ends are accurately registered. To enable this stacking operation to be conveniently carried out, the layers used are normally consolidated along their entire length by an adhesive which is

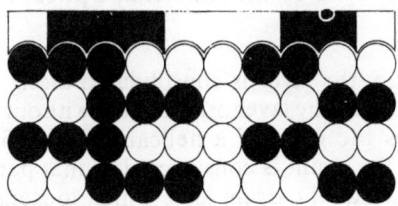

Fig. 8   Diagram illustrating the stagger introduced into the positions of fibres in a coherent bundle wound using the coil-winding technique.

Fig. 9   Photograph of an image transferred through a bundle made by the coil winding technique, showing poor coherence in the laying direction, (vertical).

dissolved out after assembly. When the required number of layers have been stacked together the ends are bonded together with a suitable resin and the bundle immersed sequentially in baths of solvent until the consolidating adhesive is removed.

This is a simple technique, its main disadvantage being that the structure of the fibre layers is such that motion of one layer relative to the next is difficult in a direction at right angles to the fibres. This can prevent the accurate registration of layers in the assembly jig with a resulting loss of coherence.

This disadvantage is overcome neatly by an elegant modification of the technique developed by the American Optical Company.

C. Hoop Technique

In this technique, the layer is consolidated along part of its length and removed by sliding the layer off the drum without slitting the fibres, so the layer takes the form of a helical hoop of fibre. To ease the removal process, the drum is constructed so that part of its periphery can be moved towards the drum axis, which loosens the layer on the drum (see Fig. 10). The consolidated portions of the hoops are then stacked into a jig, each hoop being bonded to the previous one at the

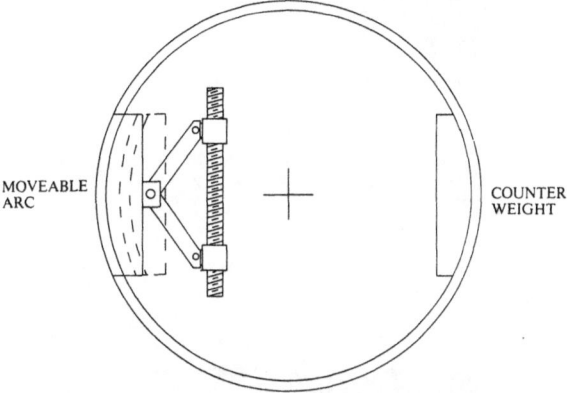

MOVEABLE
ARC

COUNTER
WEIGHT

Fig. 10 Drum design used in the manufacture of coherent layers to be used in the 'hoop' technique. The moveable sector is used to loosen the wound layer on the drum, after winding is complete.

consolidated region during this stacking process. It will be appreciated that any misalignment of the hoops in this process merely shifts the entire layer laterally and will not misalign the fibre ends in the completed bundle since these ends are still joined together. Once the required number of hoops are bonded together, the assembly is slit through the bonded region. Obviously, some indication is required on each hoop to indicate the direction of the helix angle so that the hoops may be stacked with these directions pointing the same way.

Some misalignment still occurs, owing to the fibres in separate layers not being parallel to one another. Part of the bonded region is removed during slitting and subsequent grinding and polishing, and fibres in the ends will be displaced relatively if they are not accurately parallel. However, this misalignment is small and the hoop technique is the best available technique for the manufacture of coherent bundles.

IV. PROPERTIES OF FLEXIBLE COHERENT BUNDLES

There are two types of coherent bundle which can be manufactured, characterized by the type of fibre (viz single or multiple fibre) used in the assembly (Fig. 11). As the choice of fibre defines certain other properties of the bundle, a comparison between the two types is useful.

In any of the manufacturing processes there is a lower limit to the size of fibre which can be handled. At present this is around 15 μm – 20 μm, and so any monofilament bundle will have a resolution capa-

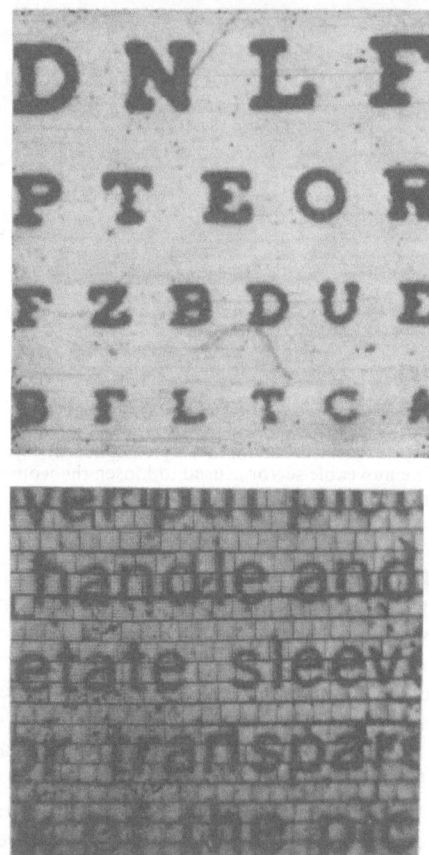

Fig. 11 Photomicrograph of mono-filament and multiple fibre coherent bundles:
(a) Mono-filament bundle fibre size 20 μm. (Courtesy of Schott and Gen.)
(b) Multiple fibre bundle fibre size 10 μm.

bility lower than that of a corresponding multiple-fibre bundle. However, monofilament bundles are preferred in a number of applications, for the following reasons.

The existing state of the art in multiple fibre bundles is such that the smallest cross-section which can be produced is around 60 μm. Typically, this will be a square multi-fibre comprising a 6×6 matrix of 10 μm fibres (Fig. 12). A single breakage in such a bundle creates a black area 60 μm square, containing 36 information points, whilst a breakage in a monofilament bundle creates a black spot, at most 20 μm in size. The presence of such black areas in the final image plane is

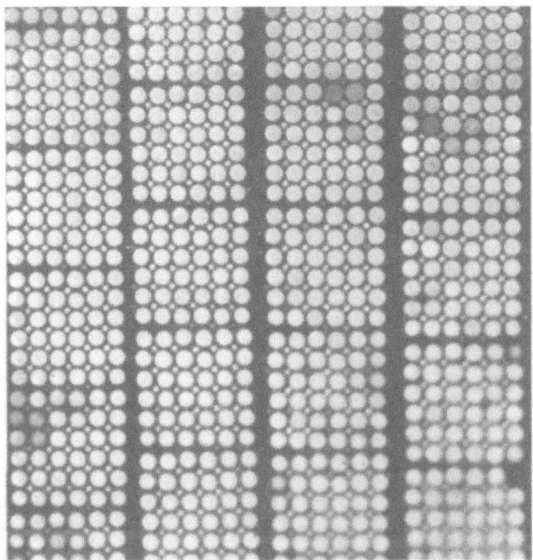

Fig. 12   Photomicrograph of 6 × 6 multiple fibre used in flexible coherent bundles.

distracting to the viewer and one relatively large area is much more distracting than a number of small areas even though the total areas may be equal (Fig. 13). In applications where the viewing conditions are critical, the monofilament bundle is therefore to be preferred; this applies particularly to the case of medical instrumentation. Further, the monofilament bundle tends to be more flexible than the multiple-fibre one, since the basic mechanical unit is much smaller and the fibre cross-section, being circular, allows the fibres to slide more easily over one another during flexing. Thus, monofilament bundles tend to be used in applications where extreme flexibility is required or where optimum viewing conditions are essential.

The two main disadvantages of the monofilament bundle are first, that the basic resolution is limited and secondly, that the manufacturing process necessarily takes much longer than is the case with multiple fibres and therefore the cost is normally much higher. Another consequence of this is that these bundles are normally supplied in smaller cross-sections than multiple fibre bundles.

Conversely, multiple-fibre bundles find their applications where high resolution is required, or where there is a need to transmit a large amount of information. Multiple-fibre bundles have been made with cross-sections of up to 50 mm, i.e. with a practical information content in excess of 4×10^6 bits.

Fig. 13   Illustration of the reduction of image quality due to broken fibres:
(a) Original photograph. (Courtesy of Barr and Stroud Ltd)
(b) Original with opaque regions corresponding to broken mono-filaments (20μ) in a 3 mm × 3 mm bundle.
(c) Original with opaque regions corresponding to broken multiple fibres ($60 \mu \times 60 \mu$) in a 3 mm × 3 mm bundle.
The total area of opaque regions is equal in (b) and (c).

A. Optical Properties

The two main optical properties of coherent bundles which are of interest are transmission and resolution.

As will be appreciated from the earlier chapters, the shape of the spectral transmission curve is determined by the basic transmission of the core glass. The actual transmission at any particular wavelength is controlled not only by the core glass, but by the core packing fraction, reflection and end-losses. The overall transmission of a coherent bundle is less than that of a non-coherent bundle, made with the same glasses, for two reasons. First, the requirement for accurate positioning of the fibres in the coherent bundle inevitably means that the fibres cannot be packed as closely together as they would be in a non-coherent bundle. Secondly, the sheath thickness is proportionately greater in the fibres used in the coherent bundle, to reduce cross-talk between individual fibres. Both of these reasons lead to a reduction in the core packing fraction. Transmission curves for typical coherent bundles are shown in Fig. 14.

The resolution of monofilament bundles is controlled, as in all fibre optics components, by the fibre diameter. The present technology can yield resolutions of up to 25 line pairs/mm in bundles up to 10 mm in cross-section. The aesthetic appearance of the transferred image is good, the only common fault being a certain fuzziness seen in straight edges, owing to small variations in fibre position. Multiple-fibre coherent bundles can yield resolutions of up to 50 lines pairs/mm with bundle sizes up to 50 mm in cross-section. Variations in fibre position cause a more serious degradation in image quality in this type of bundle, in which a straight edge appears "castellated" (see Fig. 15).

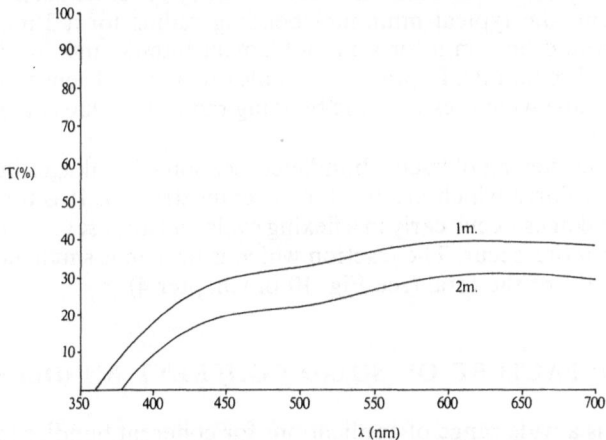

Fig. 14   Spectral transmission of coherent bundles 1 m and 2 m long.

Fig. 15 Photograph of a line transferred through a multiple fibre coherent bundle, showing the castellation effect.

B. Mechanical Properties

Like all fibre optics components, coherent bundles are very robust when mounted correctly and can withstand high acceleration forces and extreme vibrational conditions. The main area of weakness is the point where the consolidated and unconsolidated regions of the bundle meet. This should be protected and is normally achieved by ensuring that this region is within the metal ferrule covering the bundle ends. The flexibility of these bundles is controlled by the cross-section of the bundle and the typical minimum bending radius for a 3 mm cross-section would be 7 mm for a monofilament bundle and 20 mm for a multiple-fibre bundle. In practice, bundles are normally encased inside a flexible tube which restricts the bending radius to values in excess of the above.

Repeated flexing of such a bundle causes some breakages but these are due to fibres which are weak or heavily stressed. It is found that these breakages occur early in a flexing cycle and after several hundred flexes no more occur. The fraction which is broken is small, normally less than 1% of the total (see Fig. 10 of Chapter 4).

V. MANUFACTURE OF SOLID COHERENT BUNDLES

There is a wide range of applications for coherent bundles in which there is no flexibility required, or the environmental conditions in use or in further processing of the bundle do not permit the use of resin for bonding. For example, some types of fibre bundle are incorporated into the envelope of vacuum tube devices which have to be baked, at the annealing point of the glass, during outgassing of the assembled device. In such cases a solid coherent bundle is used in which the fibres are bonded together by the fusion of their sheaths, as in a multiple-fibre. In fact, the manufacturing process is a repetition, and modification, of the multiple-fibre drawing process which has already been described. A convenient starting point for this process is indeed a multiple fibre. For ease of handling, the multiple fibres used in solid bundles have a much larger cross-section than that required by flexible bundles. Typical values of cross-section range from 1 mm to 3 mm, and the number of individual fibres will normally be much larger than the 36 fibres used for flexible bundles.

Typically, a multiple fibre for a solid bundle would be produced by stacking 1.0 mm fibres into a square or hexagonal format which might be 40 mm across flats, and would contain around 1000 individual fibres. This stack could be drawn to produce a multiple fibre with a cross-section of a few millimetres. If a very small fibre size is required the drawing process can be repeated, using stacks of multiple fibres, to produce a two-stage multiple fibre. In this way, multiple fibres can be produced which contain 5 μm fibres and have a cross-section of 3 mm. It will be appreciated that these multiple fibres constitute image-transferring devices with performances at the very least comparable to those of flexible coherent bundles. Indeed such components are used for the transmission of images and can be bent to complex shapes in a flame to suit any desired light path. A multiple fibre used in such a fashion is commonly called an image conduit.

The two-stage drawing process is intrinsically much more difficult than a single-stage process since the original multiple fibres are softened again in the second stage and any air trapped between the fibres during the first stage can expand in the softened glass and severely distort the fibres. In addition, any distortion caused by the fusing together of the fibres in the first stage is normally small compared to the fibre size. This is not so in the second stage, since the individual fibres are very much smaller, and to minimize this the cross-sectional sizes of the initial multiple fibres must be closely controlled and the stacking of the second-stage bundle must be very accurate.

A number of techniques can be used to ensure close packing of the multiple fibres during the second-stage draw and perhaps the most widely used is shown in Fig. 16. In this technique the stack is cemented on to a holder with provision for drawing a vacuum through the fibres. The stack is wrapped in a thin metal foil almost down to the working zone of the furnace and on evacuation this is drawn tightly on to the

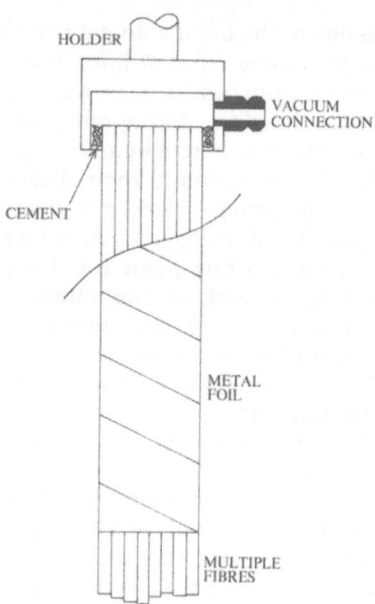

Fig. 16 Diagram showing the use of metal foil to maintain good packing in a second stage, multiple fibre drawing process.

stack, ensuring good packing along its length. The foil is unwrapped from the bottom of the stack throughout the run, as it enters the furnace. Obviously, great care must be taken to exclude refractory material from any of these stacks. Since 1.0 μm particles can cause serious distortion of a 5 μm fibre, it is essential that the drawing of such fibres be done in an area in which the atmosphere is filtered.

The above process can yield multiple fibres of cross-section up to about 6 mm, depending upon the final fibre size. However, several important applications demand a cross-section of 20 mm to 150 mm. This cannot be achieved by an extension of the drawing process and bundles of such cross-section are made by the fusing together of stacks of multiple fibres, without a corresponding reduction in diameter. There are probably as many ways of doing this as there are fibre-optics manufacturers, but the following two techniques serve to illustrate the basic principles.

A. Fusion Using Fluid Pressure

This technique is used in the large-scale manufacture of the smaller cross-section bundles, up to 50 mm, and the basic equipment is illustrated in Fig. 17. The multiple fibres are closely packed into a glass tube, closed at one end, whose bore corresponds to the bundle cross-section; this glass should have similar thermal characteristics to the fibres. This tube is then affixed to a vacuum connection and mounted, as shown, vertically inside a muffle furnace. This furnace has two elements, a main element and a small moveable element which can travel up the muffle. The combination is taken up to the annealing temperature of the glass by the main element of the furnace. At this stage oxygen is bled into the tube to oxidize any carbonaceous material that may be present, and the tube is evacuated and outgassed for some time.

After outgassing, the vacuum is maintained and the small furnace element is slowly moved up the muffle starting at the bottom of the tube. The temperature of this element is such that the zone of the tube encircled by the element is taken up to a temperature which causes the fibres to soften and be fused together by the collapse of the glass tube owing to the differential pressure created by the vacuum. By a suitable choice of temperature and rate of ascent, the multiple fibres can be fused together so that no interstitial voids exist and a slice taken from the fused boule will be vacuum-tight. When the moveable element has reached the top of the fibres, the power to this element is switched off and the fused boule left to settle down to the annealing temperature. The entire boule is then annealed by cooling the furnace at a predetermined rate, using a programme-controller on its main element.

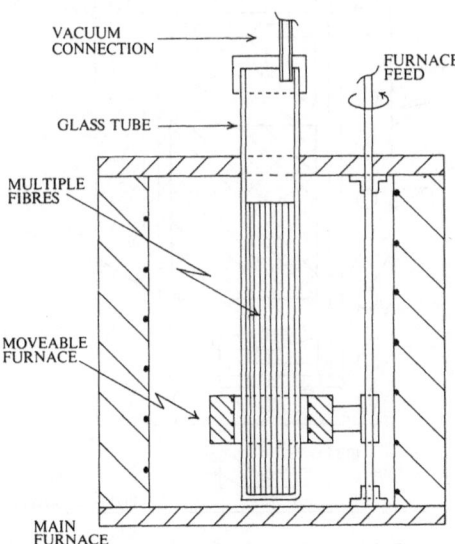

Fig. 17   Basic layout of the equipment used in fluid pressure fusion.

In one modification of this technique, the outside of the glass tube is pressurized to increase the pressure differential. Using this technique fused boules can be produced up to 50 mm in diameter, and up to 1 m long. The boule can be cut into solid bundles of the required length for final grinding and polishing.

B. Fusion Using Mechanical Pressure

This technique is used to produce small quantities of larger cross-sections, normally up to 130 mm. In this case, the multiple fibres are stacked in a metal former which defines the cross-section of the final bundle. To prevent the fibres from sticking to the sides of the former, a thin layer of a release material, e.g. mica, is interposed between the fibres and the former. The former is mounted in a pressure bomb in a mechanical press as shown in Fig. 18.

The metal former fits closely into a pressure bomb which has two moveable plattens top and bottom by means of which pressure can be applied to the fibre stack. To enable the interior of the bomb to be evacuated, stainless steel bellows are welded between the plattens and

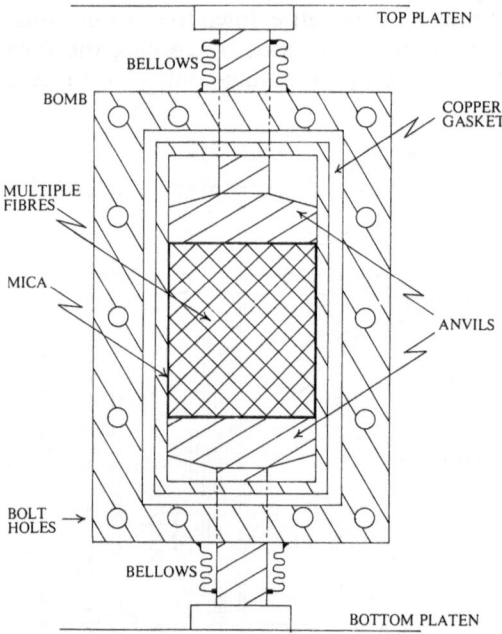

Fig. 18   Basic layout of the bomb, with front plate removed, used in fusion with mechanical pressure. The bolt holes carry the fixing bolts for the front plate.

the main body of the bomb. The bomb is finally sealed by bolting on end-pieces, via copper gaskets. The bomb is heated up by a surrounding furnace to the annealing temperature where oxygen is bled into the bomb and the whole system is outgassed. The entire bomb is then taken up to the fusing temperature and pressure is applied to the fibres, while maintaining the vacuum. After the fusion is complete, the bomb is taken down to the annealing temperature of the glass and held there until the whole mass is at this temperature. The fused boule is then cooled, through an annealing cycle, to room temperature when it can be removed from the bomb and processed.

The pressures used are much higher than those required in the former technique; the reasons for this are not yet fully understood but appear to stem from two causes. First, the stack is compressed by about 10% of its original height during fusion and there will be large viscous forces to overcome in doing this. The effect of these forces can be seen in bundles made by this technique, where rows of multiple fibres which were originally horizontal in the bomb are curved along the direction of fusion. In addition, there is normally evidence of distortion in the multiple fibre cross-section (Fig. 19). Secondly, and probably more important, the pressure is applied over the whole length of the multiple fibre so that it is possible for a channel of air in the interstices to become sealed at both ends. If a vacuum-tight bundle is required, one

Fig. 19   Photomicrograph of fused plate made by mechanical fusion, showing distortion in shape of the square multiple fibres. The final diagonal ratio is 1.1:1. (Courtesy of Barr and Stroud Ltd).

Fibre Optics

must use sufficient pressure to either dissolve the gas in the glass or break the channel up into separated bubbles. This does not occur in the former technique which, by virtue of its "milking" action, tends to squeeze any gas upwards to the unfused portion of the stack.

C. Design of Solid Bundles

Because of the complex processes involved in the manufacture of these solid bundles, there are a large number of ways of achieving the desired end result. Also, these solid bundles are invariably used in situations where the bundle length is much less than the cross-section. In fact these bundles have become known as fused plates, and the bundle length (or plate thickness) is typically about one-tenth of the cross-sectional dimension (Fig. 20). Because of this, the use of e.m.a. is normal in these bundles. The most suitable design of a bundle alters as the fibre diameter decreases at a diameter of about 15μm since two important changes occur at or about this diameter. First, this diameter corresponds to the lower practical limit of a single-stage multiple fibre, and bundles which require fibre diameters less than this almost invariably use a double-stage multiple fibre. Secondly, the type of e.m.a. used is normally interstitial below this diameter and concentric above.

Above this diameter of 15 μm, the bundle will consist of single-draw multiple fibres with concentric e.m.a. It will be recalled from Fig. 4 that concentric e.m.a. is essential in the larger fibres, since the prob-

Fig. 20 Photograph showing the variety of fused plates used in vacuum tubes. (Courtesy of Barr and Stroud Ltd)

ability of unwanted light intercepting a rod of interstitial e.m.a. decreases as the fibre diameter increases. Also, these multiple fibres are normally hexagonal in cross-section since this gives better stacking. Below a diameter of 15 μm, the bundle will tend to have double-stage multiple fibres with interstitial e.m.a. These multiple fibres will tend to be square since this is the best shape for the second-stage draw, that is, square multiple fibres can be stacked to form a perfectly square cross-section for the second drawing stage.

There are two different approaches to the use of interstitial e.m.a. In the first, every available interstice is filled with a rod of the e.m.a. glass at the first multiple-fibre stage and this means that there is no e.m.a. in the interstices formed by the joining of two multiple fibres (see Fig. 21); however this does not significantly affect the performance of the finished bundle. In second approach, the e.m.a. occupies only a fraction of the interstices and can therefore only be used with very small fibre diameters, e.g. less than 7 μm. The normal method of achieving this is to draw an initial multiple fibre which has only four fibres in a square format, and the e.m.a. rod occupies the central interstice. This multiple fibre is then stacked and re-drawn twice to form the final unit, as illustrated in Fig. 22.

The use of more than one multiple-fibre drawing process has advantages which have prevented the development of single-stage multiple fibres for small fibre diameters. In the single-stage multiple fibre more individual units (i.e. single fibres) have to be handled than in a double-stage, or triple-stage, multiple fibre, and the surface area exposed to the atmosphere throughout the process is significantly increased. To illustrate this, the number of units and relative total surface for unit area of bundle have been calculated for the case of a square multiple fibre with 5 μm fibres. It is assumed in each case that the fibre stack is built into a 60 mm square former.

In the case of a single-stage multiple fibre the single fibre diameter

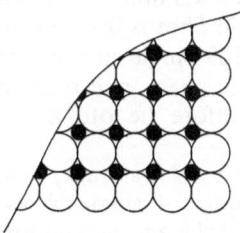

Fig. 21   Diagram showing the position of interstitial rods, (black circles) in a multiple fibre stack.

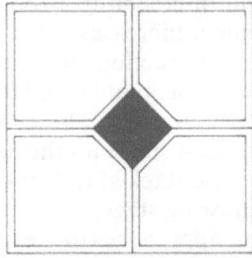

Fig. 22 (a) Diagram showing the position of central interstitial rod in the first stage
of a three-stage draw, used for final fibre diameters below 7 μ.
(b) Photomicrograph of a fused plate made using this technique. Fibre size is
7 μ. (Courtesy of Schott and Gen.)

required for the minimum number of units is about 1.5 mm. Thus, there
will be some 1600 single fibres in the multiple fibre which will have a
final size of 200 μm. There will therefore be 2500 multiple fibres per
unit area and the total number of units required is 4100.

In a two-stage process, the minimum number of units can be shown
to be obtained when the various fibre sizes are approximately as
follows. The single fibre is 4.5 mm, the first-stage multiple fibre is 4.0
mm and the final multiple fibre is 0.8 mm. In this case the first-stage
multiple fibre requires 169 single fibres. The second-stage multiple
fibre requires 225 multiple fibres, and there will be 125 final multiple
fibres per unit area. Therefore the total number of units which has to
be handled is 519.

In a triple stage, with a 2 × 2 first stage, it can be shown that the first-
stage multiple fibres require 4 single fibres, the second stage, 100
multiple fibres and the third stage, 400 multiple fibres. The density of
final multiple fibres is 100 per unit area. The total number of units
required is therefore 604. This latter figure is higher than with the two-

stage process since the initial restriction placed on the first stage does not permit optimization of the number of units required. The surface area exposed for a given volume of material in a single-stage process is therefore approximately double that in the other processes. Thus, from a handling point of view, the use of more than one multiple-fibre-drawing stage is very attractive.

In some specialized applications, a very high degree of alignment is required of the fibres. Whilst the act of drawing a multiple fibre ensures such alignment within the multiple fibre, the process of fusion can cause some misalignment of fibres in adjacent multiple fibres. This can be overcome by using multiple fibre cross-sections which are interlocking, and two such cross-sections are illustrated in Fig. 23. Obviously the dimensions of this type of multiple fibre must be more accurately controlled than is necessary with square or hexagonal cross-sections, and the cost of such bundles is correspondingly higher.

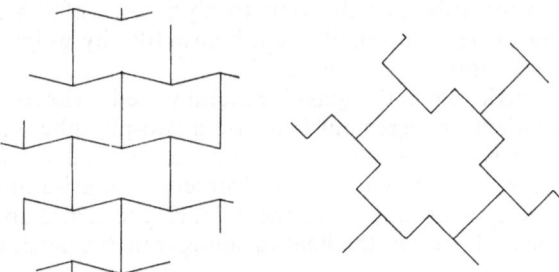

Fig. 23   Illustration of two types of interlocking, multiple fibre cross-section.

VI. EXTRA-MURAL ABSORPTION

As was noted earlier, the amount of stray light which can reach the final image plane is objectionable in certain instances and this can be significantly reduced by incorporating an absorbing material in the bundle, external to the fibre. Since the presence of this material reduces the core-packing fraction it is desirable that the volume absorptivity be as high as possible. This would suggest the use of metallic substances but there are two reasons why this is not done. First, a metallic coating cannot be drawn down in a similar fashion to glass, owing to its rapid change of state at the melting point. Thus a multiple-fibre drawing process would be impossible and any bundle would have to be made up from single fibres of the required size, which had been coated with the metal. Whilst this might be a suitable technique with fibre sizes of about 20 μm, it becomes impractical below this size. Secondly, many applications for fused plates require a high degree of electrical insulation across the faces of the bundle which is impossible to achieve if a metallic layer is used.

Because of these practical points, the most widely used e.m.a. material is a doped glass, although ceramic materials have been evaluated. The basic glass material is normally similar to the sheathing glass to ensure thermal compatibility and is rendered absorbing by the incorporation of a small percentage of metallic ions. The absorption of a metallic oxide is strongly wavelength-dependent, so that a mixture of oxides is normally used to give an approximately constant absorption over the visible wavelength band. A suitable e.m.a. glass can be made using the correct proportions of the oxides of cobalt, manganese and nickel, whose absorptivity curves are shown in Fig. 24. The ferric ion is also a very effective absorbing material, but, unfortunately, it is difficult to maintain this oxidized state in some glasses. These metallic oxides are not glass-formers and their inclusion in the glass increases any tendency towards devitrification. Partly because of this the fraction of colouring oxides is normally kept below 10% by weight, although in special cases this can be increased to 15%.

For concentric e.m.a. the glass is generally used in the form of a tube and a convenient arrangement is to use a two-ply tube, in which the inner cylinder is of clear glass and forms the normal sheath, and the outer cylinder is the e.m.a. sheath. Problems can arise in the manufacture of this type of tube, since the colouring ions also absorb infrared radiation and prevent the heat escaping from the inside of the tube

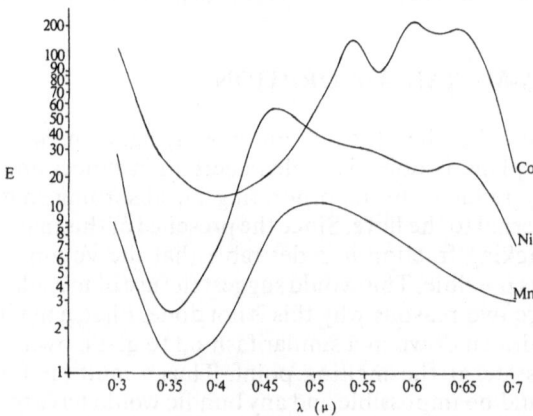

Fig. 24 Spectral variation of molar absorption coefficient (E) for the metals, nickel, manganese and cobalt in sodium silicate glasses with the approximate composition (40% Na_2O, 60% SiO_2). E is the absorption coefficient (cm^{-1}) divided by the molar density (gms/ l) (after Bamford[2]).

during manufacture. In addition, the outer surface is also a very efficient radiator and cools much more quickly than a clear glass. Both of these effects lead to a high degree of strain in the manufactured tube, which requires careful annealing before drawing. In fact, it is this which limits the amount of colouring oxide which can be added to the base glass rather than any onset of instability. If a higher degree of absorption is required, the glass can be cast into slabs which are drawn down to form rectangular strips, say 2 mm × 0.5 mm, which can be cemented on the outer surface of a tube of clear sheath glass to form an almost continuous annulus. This technique permits a much higher concentration of colouring oxide, which is then limited only by the stability of the resulting glass.

With interstitial e.m.a. the glass is made in the form of small rods, which are drawn down to the correct diameter; since the manufacturing conditions are not as critical as for two-ply tubing, it is glass instability that determines the concentration limit for the colouring oxides.

With very fine fibres, problems can be encountered owing to diffusion of the colouring ions, during the drawing process, into the clear sheath or, indeed, into the core itself. The cobalt ion is particularly mobile and most plates made using this ion in the e.m.a. glass exhibit a blue tinge in the transmitted light especially at high angles of incidence. Diffusion can be reduced to negligible proportions in glasses where the ferric ion is stable by incorporating titanium into the glass, since the titanium and ferric ions form a complex which is relatively immobile. The glasses used for this type of e.m.a. are invariably boro-silicates.

VII. PROPERTIES OF SOLID COHERENT BUNDLES

A. Optical Properties – Transmission

The most widely used form of solid coherent bundle is the fused plate. This is normally incorporated as the front screen in a photoelectronic vacuum tube where the source of light is a phosphor deposited on the inside face of the plate. The phosphor behaves as a Lambertian source and any test procedures should be designed with this fact in mind. Generally the transmission is measured using both collimated light and diffuse light. These two measurements give different information about the bundle. In the former, the transmission of the plate is measured using a collimated beam, from a monochromator, which is incident normal to the bundle face. This measurement gives information about the spectral response of the core glass for all bundles, and gives a measure of the core packing fraction for bundles, with e.m.a.; this stems from the fact that only light entering the bundle through the core in such plates will be transmitted.

The Lambertian transmission is measured using a diffuse source and a detector immersed on to the plate. This does not yield much useful information unless the bundle incorporates e.m.a., when the transmission is a measure of the effective N.A. of the fibres. This can be lower than the theoretical figure for two reasons. First, the clear sheath might be too thin so that some of the accepted light in the cores will be slightly absorbed at each reflection. Since the number of reflections and the depth of penetration into the sheath increases with increasing incidence angles, this absorption will be greater for the higher angles. This has the effect of clipping the outer edges of the polar response, giving an effective N.A. of less than the theoretical. Secondly, some of the colouring ions may have diffused into the clear sheath with effects similar to that discussed above.

B. Optical Properties – Resolution

The resolution of solid coherent bundles is measured using M.T.F. techniques and there are two basic ways of obtaining results. The first of these is the straightforward method of measuring the depth of modulation in the transmitted image of a series of bar charts, a typical set-up being shown in Fig. 25. These measurements can again be taken with collimated and diffuse sources where the former yields information about the fibre size and packing, and the latter, about the effectiveness of the e.m.a. used. Typical responses are shown in Fig. 26.

In the second method, a photometric trace is taken across the transmitted image of a straight-edge using a diffuse source (Fig. 27). It can be shown that such a trace contains all the information necessary to obtain the M.T.F. response. However, the calculations involved in this are laborious and, since the shape of this trace is in itself a useful indication of the bundle's performance, the results of such a measure-

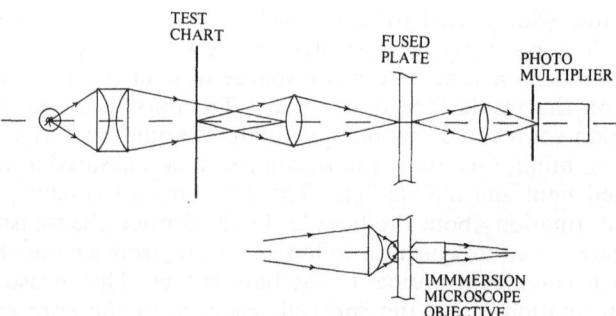

Fig. 25   Layout of apparatus for contrast measurement in coherent bundles. The upper section is representative of low N.A. measurements, the lower section shows the arrangement, using immersed lenses at the bundle, for high N.A. measurements.

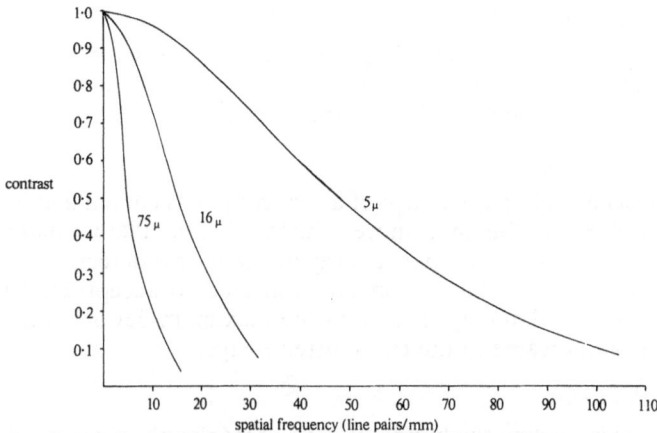

Fig. 26 Typical contrast curves obtained from fused plates with e.m.a.

ment are normally expressed as a measured response at specified distances from the position of the straight-edge. This second method has obvious attractions when measuring plates with one, or both, faces curved, since a much smaller area of the surface is used in taking the measurements.

C. Optical Properties – Defects

There are a number of defects which can occur in a fused plate, and these can be defined as follows:

Blemishes

These are black areas, normally defined as being less than 50% transmission, which are caused by the distortion of fibres by entrapped bubbles, dirt, etc. In a specification the number of allowable blemishes will be stated as a maximum areal density for a given blemish size.

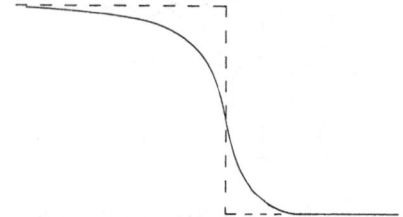

Fig. 27 Variation in light output obtained by scanning the transferred image of a straight edge. The position of the straight edge is shown by the chain line.

Shear Distortion

This is caused by the twisting of a multiple fibre at some stage in the manufacturing process, which means that the image of a straight line will be discontinuous as illustrated in Fig. 28.

Gross Distortion

This occurs when the image of a straight line is curved and is due to uneven flow of the multi-fibres during fusion. Plates made using mechanical pressure are particularly prone to this defect.

Gross and Shear Distortion are restricted to acceptable limits in manufactured plates by specifying the maximum deviation from the initial line allowable in the transmitted image.

Frame Run-Out

This occurs when the bundle is cut so that the fibres are not normal to the end-faces. This means that the whole image is moved transversely relative to the input.

All these defects occur to a greater or lesser extent in all solid bundles and must be borne in mind when designing any equipment which will incorporate such a bundle.

D. Mechanical Properties

Strength

From a mechanical point of view a solid bundle can be treated as a solid piece of glass and can be worked using standard glass grinding and polishing equipment. Perhaps the only difference observed is a certain anisotropy in the formation of chips. The mechanical strength of the bundle is less than that of an equivalent plate-glass block; measured samples indicate a strength of about half that of plate glass.

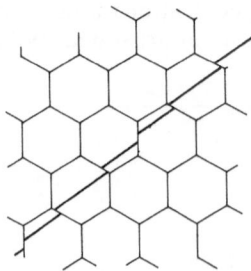

Fig. 28 Illustration of the transfer of a line through a coherent bundle which has shear distortion.

Vacuum Tightness

The fused-plate type of bundle finds its major application as the front screen in a vacuum tube device, e.g. C.R.T.s, and must therefore be vacuum-tight. This is normally checked by means of a leak detector using helium as the search gas with a mass-spectrometer detector tuned to helium, such detectors can register leaks of 10^{-12} cm^3/s at a pressure differential of one atmosphere. If one applies classical viscosity theory to the problem of helium passing through a channel of radius a and length l with a pressure differential of one atmosphere, then one finds that the leak rate if given approximately by:

$$v = 10^9 \frac{a^4}{l}$$

If there are n such channels in a fused plate, the total leak rate is given by:

$$V = 10^9 \frac{na^4}{l}$$

where the dimensions are in cm.

In order to pass the helium leak test, the number of channels is related to the channel diameter by the following inequality:

$$na^4 < 10^{-21} \tag{2}$$

where it is assumed that l is of the order of 1.0 cm. So that even if there were only one channel, its diameter must be less than 10^{-5} cm if the whole plate is to pass the test. This is obviously a stringent requirement when one considers that some plates have a fibre density of 10^6 per square cm. Some manufacturers therefore adopt sealing techniques to render plates vacuum-tight. In one such technique, the plate is immersed, under vacuum, in tetra-ethyl orthosilicate, which fills any channel. The plate is afterwards baked at a temperature sufficient to break down this compound into silica which blocks the channels. This blockage may be only effective along a small part of the channel length and some manufacturers' plates have a dielectric strength no greater than that of air, and this must be borne in mind when considering the use of such plates in CRTs where the front screen is held at a high potential relative to earth.

REFERENCES

1. H. H. Hopkins and N. S. Kapany, *Nature* **173**, 39 (1954)
2. C. R. Bamford, Physics and Chemistry of Glasses, **3**, 189, (1962)

CHAPTER 8

Coherent Bundles – Applications

The manufacture of coherent bundles is a much more difficult process than that of non-coherent bundles and therefore the range of components which can be made is much smaller than with non-coherent bundles. In fact, at the present time, the vast majority of applications involve the use of image-transferring bundles, which can be used either for direct viewing by the eye, or the transferred image can be processed electronically or photographically. For the sake of convenience, these two categories of use will be considered in two separate sections and a third section will deal with applications which do not use the conventional design of bundle.

I. DIRECT-VISION APPLICATIONS—FLEXIBLE COHERENT BUNDLES

As was mentioned above, the output face of the bundle in this range of applications is viewed directly by the eye. One consequence of this is that, unless a very high viewing angle is required, the fibres comprising the bundle can have a relatively low N.A., as low as 0.3 in some cases. In order to maintain the fidelity of colours in the image these bundles are normally made using a core glass with a relatively constant absorption in the visible part of the spectrum, although for lengths in excess of 1.0 m the transferred image still has a yellowish tinge. This is normally acceptable under all but the most stringent of viewing conditions.

In these applications the user requires to view an area which is remote and normally inaccessible by conventional techniques, e.g. a rigid optical telescope. In the fibre optics instrument, the area to be examined is imaged on to the remote end of the bundle by means of a suitably positioned objective lens which gives the required degree of magnification. The output is viewed through an eyepiece which magnifies the transferred image to the optimum size. Figure 1 illustrates schematically this arrangement. It will be realized that this type of application comes under the heading of inspection, and is to be found mainly in the engineering and medical fields.

Instrument Design

The objective lens is designed so that the cone of light falling on each fibre is within the acceptance angle of the fibres, which ensures

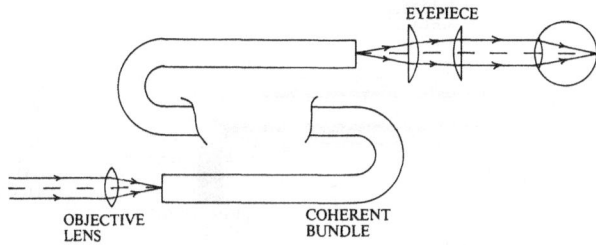

Fig. 1   Basic layout of the optical system for a coherent bundle used as an inspection device.

that the amount of stray light due to rays not being trapped by the fibres is minimized. However, the acceptance angle of the fibres should not be so large that light which is not part of the required image is accepted by the fibres. This stray light can arrive at the bundle face by reflection from the lens mount or other metal parts. The lens design can normally be relatively simple, a high performance not being necessary, since the resolution in the final image is restricted by the coherent bundle, and a satisfactory lens need only have a high contrast up to, say, 50 line pairs/mm, for present-day bundles. In addition, the effects of field curvature can be allowed for in the total design by curving the input face of the bundle to match the image plane of the lens. Due consideration must be given to the degree of magnification required with this lens, bearing in mind the limited resolution of the bundle. For example, if one has a bundle with a resolution of 50 line pairs/mm and an objective lens position which yields a reduction of 10:1 then the resolution at the object plane will be 5 line pairs/mm.

The eyepiece is normally of a standard design; the only point to note is that the magnification has an upper limit which is determined by the size of the fibres, since it is pointless to magnify the image so that individual fibres are clearly discernible. If this were done the image would be overlaid with a mosaic pattern of the fibre ends, which is distracting. With current fibre sizes (10 μm–20 μm), the eyepiece magnification should not be in excess of × 10.

A typical design for such an instrument is shown in Fig. 2 where the coherent bundle is encased in a flexible metallic trunking for protection, which also restricts the bending radius to a safe figure. It will be seen that a light guide is incorporated into this sheathing and is used to illuminate the area of interest. The output from this guide does not pass through the objective lens, since this would cause the illuminating light to be focussed on the object plane outside the field of view of the coherent bundle. In addition, Fresnel reflection from the objective lens elements would seriously degrade the contrast. If the path over

Fibre Optics

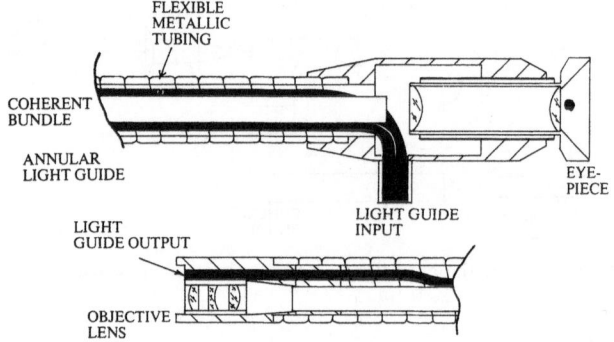

Fig. 2   Mechanical layout of a flexible inspection instrument incorporating a light guide for illumination.

which the instrument has to travel is restricted, the light guide can be omitted from the design in order to reduce the overall cross-section. The light required for illumination is provided from a filament lamp whose output is focussed on to the end of a light guide. The other end of this light guide is butted against the integral light guide by a spring clip, which provides a satisfactory means of introducing light into the instrument. In an instrument with integral illumination there is obviously an optimum size for the light guide, since, for a given instrument cross-section, the guide area must decrease as the objective-lens diameter increases. It is a simple matter to show that this optimum occurs approximately when the area of light guide end-face equals that of the entrance pupil of the objective lens. There are a number of refinements which can be added to this basic design, the two most common being remote focussing and steerability.

Remote Focussing

Some commercial instruments give satisfactory performance with a fixed objective lens, which might typically have a focal length of 5.0 mm and be positioned to give a depth of focus from about 50 mm to infinity. The aperture of this lens can be fairly large, since the range of focus is effectively determined by the fibre diameter, as illustrated in Fig. 3. It is a simple matter to show that, in the Figure, if the further of the two object positions is infinity, then the closer is given by $D/2d$ where D is the lens aperture, and d is the fibre diameter. However, at the closest object point the resolution is only about 5 line pairs/mm at the object plane and this is inadequate for many applications. Some means is therefore necessary which would allow the user to adjust the focus of the objective lens while the instrument is in position, so that an initial search can be made viewing a large area. When the area of interest has been found a detailed inspection of this can be carried out

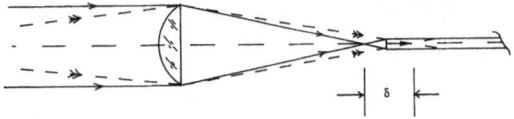

Fig. 3   Diagram illustrating the range of focus available in a coherent bundle. The fibre on the right accepts light from an infinite point and that from a closer point, indiscriminately. The closer object point is imaged at a distance, δ, from the focal point of the lens.

under higher magnifications. This remote focussing is invariably carried out by moving the end face of the coherent bundle relative to the objective lens, which is fixed. A typical method of achieving this is shown in Fig. 4 where the end of the coherent bundle is constrained to slide along the lens axis on guide rails and is attached to the inner cable of a flexible control cable. The other end of this cable is attached to a drum at the eyepiece, which can be rotated to move the remote face of the bundle relative to the objective lens, thus altering the focus.

Steerability

Since the two ends of the instrument are connected by a flexible system, it is difficult to manoeuvre the remote end accurately by the movement of the eyepiece and, in fact, if there is a restriction in the path, such manoeuvres are impossible. During a search and location procedure such manoeuvrability is essential and various means of steering the remote end have been devised. The basic concept is shown in Fig. 5 and can be seen to consist of encasing a length of the bundle at the remote end in a helical spring. This spring has a natural tendency to remain straight but can be bent into an arc by pulling the control cables. These control cables terminate at the eyepiece end in a joy-

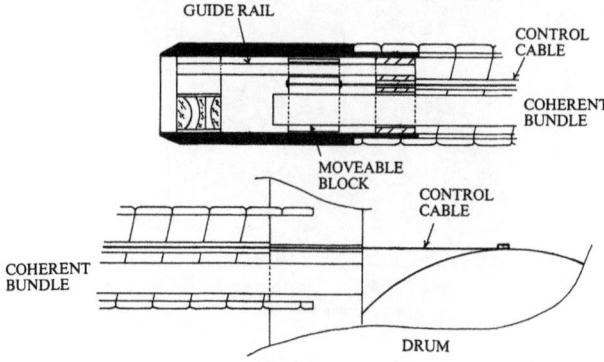

Fig. 4   Illustration of a remote focussing arrangement for a coherent bundle. The bundle is fixed to the moveable block which can be moved along the guide rail by the control cable.

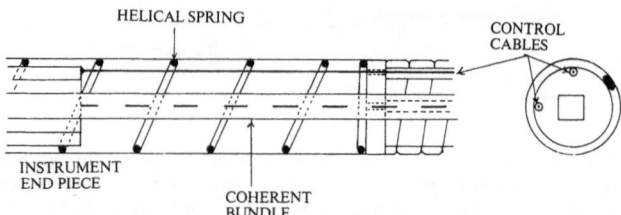

Fig. 5 Diagram showing the incorporation of a helical spring between the instrument end-piece and the trunking to provide a remote steering facility.

stick arrangement, movement of which will point the remote end in almost any direction. In a simpler version, only one cable is used which bends the spring in one plane only and the full range of movement is achieved by rotating the whole instrument about its axis.

These flexible viewing instruments or fibrescopes, as they are commonly called, provide inspection means for areas which are inaccessible by more conventional systems. As was mentioned earlier, these find their major applications in industrial and medical inspection; their use in the latter field will be dealt with in a separate chapter. An industrial fibrescope, Fig. 6, has to be rugged and this means that the flexible protective sheath is normally of convoluted metal and, in some instruments, this has a 'flex and stay' property so that the instrument can be formed to a complex shape prior to insertion into the inspection area. A number of accessories are normally provided with such an in-

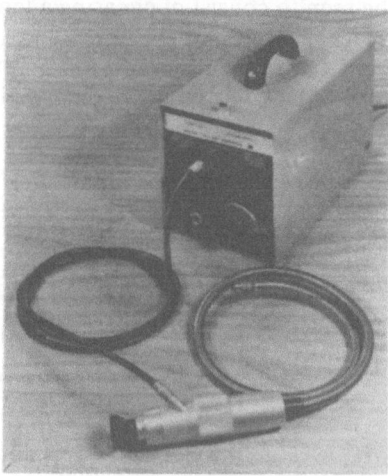

Fig. 6 An industrial fibrescope, which incorporates an illuminating light guide fed by the light source shown (Courtesy of Rank Precision Industries Ltd).

strument which may include, for example, mirror attachments to permit viewing at angles to the instrument axis and a centring attachment to maintain the instrument central in the bore of large cylinders. The range of industrial applications is wide, ranging from fairly coarse requirements such as looking for foreign objects of a reasonable size, e.g. sand deposits inside castings, to relatively fine requirements, such as looking for hairline cracks in stressed components, e.g. turbine blades. For the former types of application the simple fixed focus instrument is adequate.

Although an inaccessible area is the normal target for these instruments, a number of applications use their flexibility in another way. In these, the requirement is to view a moving or vibrating object and this is achieved by positioning the remote end of the instrument so that it is fixed with respect to the object; the viewer can then see the object as if it were stationary. This is of particular interest in the vibration testing of mechanical components.

Obviously, it is possible to take photographs of the images seen through such instruments, although they are not designed expressly for this purpose, and most commercial instruments can easily be adapted to take a camera. This should be of the reflex type so that a photograph can be taken when the desired view is obtained without the need for dismantling the eyepiece end, with the attendant risk of moving the remote end of the instrument.

Quality Control

Coherent bundles have also been used in inspection applications where widely separated areas have to be viewed simultaneously. An example of this is shown in Fig. 7 which illustrates an inspection jig for a large mechanical assembly. In such an assembly, the final inspection must ascertain that the lengths of sides, corner angles and hole positions are correct. This is done by positioning coherent bundles on the jig, each with objective lenses imaging the important features, i.e. corners and holes. The other ends of these bundles are brought together and viewed through a common eyepiece; thus all the important features are collected into a small area which can be viewed easily. The correct position for each of these features is shown on a graticule which is superimposed on the output image, and any variation in size is immediately obvious to the viewer. This type of jig can be designed using conventional optics, but this would be difficult to reset for a different shape of assembly. Resetting is straightforward using coherent bundles.

Safety Control

A rather interesting range of applications has emerged recently in which very low resolutions are used, e.g. a few line pairs/cm. These

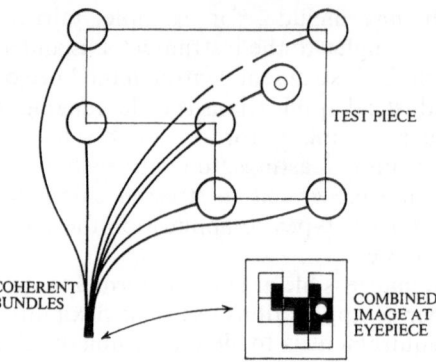

Fig. 7   Illustration of the use of coherent bundles to inspect large mechanical components.

are to be found in the area of safety controls where it is essential to know that a mechanical movement has taken place. This is done conventionally using a micro-switch, actuated by the movement, to operate an indicator circuit. However, the micro-switch circuit itself can fail and normally fails safe, so that a warning is given even though the whole system is functioning perfectly. By using a coherent bundle, direct visual evidence of movement can be provided. This type of application is increasing in use, particularly in aircraft interlock systems, e.g. checking that the undercarriage is raised or lowered. The fibre sizes in the coherent bundles used in these systems would be typically 100 μm to 150 μm.

II. DIRECT-VISION APPLICATIONS—SOLID COHERENT BUNDLES

As was discussed in the previous chapter, solid bundles can be obtained in two forms. The first of these is the image conduit in which the length of the bundle is greater than the cross-section and the second is the fused plate, in which the converse is true. Since the applications of these two forms are significantly different, they will be treated separately.

A. Direct-Vision Applications – Image Conduits

Image conduit is, in effect, a solid version of the flexible bundle and there is a certain similarity in the applications of both these components. Obviously, image conduit cannot be flexed but, if flexibility is not required, this provides a coherent bundle which is cheaper and can have a higher resolution and transmission than the flexible bundle.

Inspection

There are two areas of inspection in which image conduits are used; the first of these is where the viewing device can be pre-formed to the required shape and still allow insertion and removal. The second is in systems where the bundle can be fixed permanently in position, providing a permanent inspection facility. This type of application is clearly identical in concept to the case of flexible bundles already discussed, and most of the previous comments are applicable. The only major design difference is that, since the image conduit is fragile, it should be mounted in, say, silicone rubber to afford protection against shock.

The major disadvantage of image conduit is that blemishes are quite common owing to the relatively long length of fused fibre in the bundle. If it were not for this, image conduit would offer serious competition to conventional instruments using a lens train, since it is capable of comparable resolutions and a higher transmission. In addition, it has none of the optical defects of a train of lenses, viz distortion, ghost images and field curvature. The main objection to the use of image conduit is the loss of the three-dimensional effect, achieved in the conventional instrument by the accommodation of the eye, since the eye sees the image on the end of the bundle which is a two-dimensional area. These bundles can be used to provide an in-built inspection facility, e.g. gear-boxes, and the basic design follows the lines already laid down.

Image Transposition

Applications for image conduit are being developed which utilize the simplicity and ruggedness of this component; a simple example of this is shown in Fig. 8, where image conduit is used to transmit the

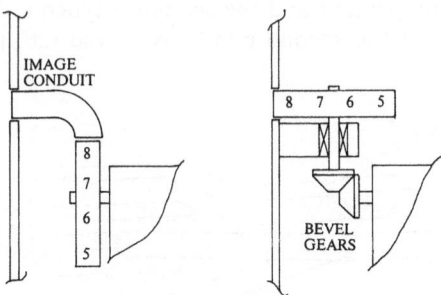

Fig. 8   Illustration of the use of image conduit (left) to transpose images. The conventional design is shown on the right.

reading on a numbered wheel to the outside of a meter case. The wheel indicates the angular position of a shaft and the conventional way of doing this is also shown in Fig. 8, using a gear train to change the position of the axis of rotation. The fibre optics solution was, in fact, the less costly alternative and two important points can be illustrated from this example. First, no lenses are used in the system; the face of the conduit is positioned directly over the numbered wheel and spaced about 0.5 mm above the surface. The N.A. of the fibres used is about 0.5 which means that the loss of contrast owing to lack of contact with the end-face is not serious for this separation. Secondly, the conduit acts as its own light source, ambient light being transmitted through the conduit to illuminate the numbers on the wheel. These two points are typical of this type of application.

A second example is illustrated in Fig. 9, in which the conduit is used to re-arrange the order of a row of numerals and is incorporated into the verifying section of a computer input terminal. In order to verify that the information being fed into a computer is accurate, the normal procedure is to have two typists type the same information on separate machines and compare their outputs. It is assumed that the probability of these operators making the same error at the same time is negligible, so that, if their outputs are identical, then the information is correct; if not, then an error exists in at least one of the outputs. A system which only required a single operator would be very useful in certain areas, but self-verification is difficult since the simple procedure of asking the operator to type the information twice is not effective, since the number sequences will be remembered unconsciously and any mistake would tend to be repeated on the second attempt.

However, by re-arranging the numbers in a random fashion, using image conduits, the sequences are broken up on the second attempt, eliminating any unconscious association. In practice two fibre bundles are positioned sequentially over the numbers to be typed; in the first there is no rearrangement and the sequence typed is the correct one. After this is typed the second bundle is moved into position, which

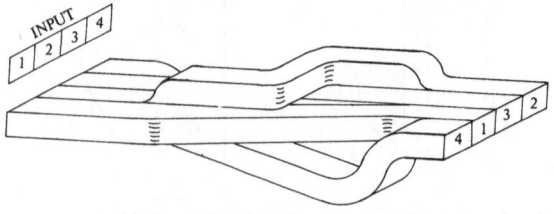

Fig. 9   Illustration of a device used to change the order of a sequence of numerals.

rearranges the order of the numerals. When this is typed, the correct order is electronically established and compared with the first attempt; if these are identical then the information has been correctly typed. In this example the first bundle is used simply to equalize the viewing distance of the numerals for both operations.

B. Direct-Vision Applications – Fused Plates

As was mentioned earlier, fused plates are solid coherent bundles in which the bundle length is smaller than the bundle cross-section. Because of this, fused plates are not used to transfer images to more convenient viewing positions, but to modify the appearance of the image in order to improve its acceptability to the eye.

High-Contrast Displays.

The widest application of these plates in this area is in a high contrast CRT display. A normal CRT displays an image or a trace as the emission of light from a phosphor screen deposited on the inside surface of the front face. This screen is a thin layer of small crystals which will scatter any ambient light falling on them and some of this light will be scattered back, giving the screen a white appearance. This decreases the contrast of any emission from the phosphor since it establishes a constant non-zero background on the screen. If this background intensity is represented by I_B and the phosphor emission intensity by I_p then the contrast is given by (see section 6.III.A):

$$C = \frac{I_p}{I_p + 2I_B} \tag{1}$$

Under conditions of high ambient lighting, this reduction in contrast can make the phosphor emission undetectable by the eye. For this reason CRTs which have to be viewed directly are provided with hoods to shield the screen from ambient lighting or must be used in conditions of low ambient lighting. This latter approach is used in radar installations and leads to rapid fatigue of the operators. An improvement in this contrast can be obtained by placing a filter in front of the screen since the ambient light will have to pass through this filter twice as opposed to the single passage of the phosphor light. This filter can be either neutral density or selected to have a high transmission in the emission spectrum of the phosphor and a low transmission elsewhere. If the effective transmission of the filter is T then the resulting contrast is given by:

$$C' = \frac{I_p}{I_p + 2TI_B} \tag{2}$$

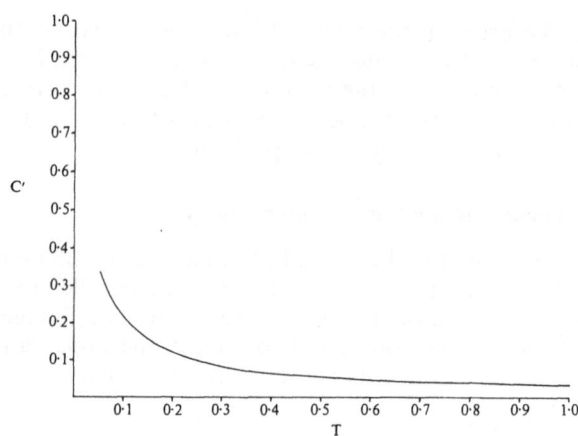

Fig. 10   The variation of contrast, C', obtained by the use of a screen in front of a *CRT*, plotted against the transmission of the screen, T.

Figure 10 shows the variation of this contrast with filter transmission for the situation in which the trace is just visible without the filter, which is generally taken to be at a contrast of 0.025. It will be seen from this that the transmission of the filter has to be less than 10% before a reasonable contrast is achieved, which means that the brightness of the trace is reduced by at least a factor of 10.

A much more satisfactory result is achieved by using a fused plate as the screen of the CRT. This plate has a low N.A. (less than 0.5) and incorporates e.m.a. Thus, ambient lighting which strikes the plate at angles outside the N.A. will be absorbed before it can reach the phosphor. The fraction of the ambient light which is within the N.A. reaches the phosphor and is scattered; this fraction is given by the square of the N.A. If the phosphor is in intimate contact with the plate, the additional background due to multiple internal reflections of scattered light at the outer surface of the CRT fact is also eliminated by the effect of the e.m.a. It can be shown that this effectively halves the background intensity from the screen. It is this secondary scattering which causes the 'halo' surrounding a bright spot on a normal CRT. Thus the contrast with a fused plate in position can be written as:

$$C'' = \frac{I_p}{I_p + (NA)^2 I_B} \tag{3}$$

Figure 11 shows the variation of this contrast with the N.A. of the fused plate, under ambient conditions similar to those adopted to produce Fig. 10 where the upper curve has been calculated taking into

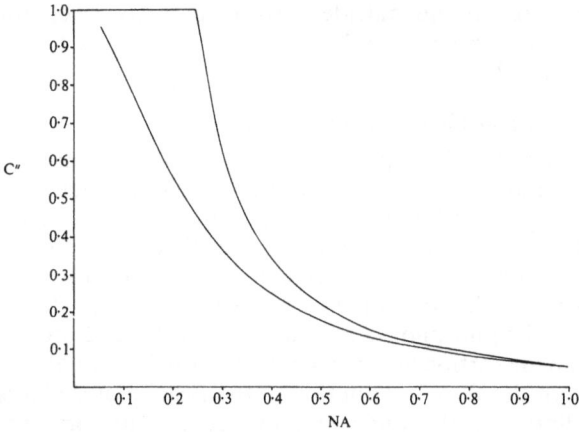

Fig. 11   The variation of contrast, C'', obtained by the use of a low N.A. fused plate, with e.m.a, as the screen of a *CRT*, plotted against the N.A. of the plate. The upper curve takes into account the obscuration afforded by the observer's head.

account the obscuration of ambient light afforded by the viewer's head. It should be remembered that in this case the trace brightness is virtually unaltered. A model demonstrating this contrast improvement, under conditions of high ambient lighting, is shown in Fig. 12. An additional benefit gained by the use of a fused plate is that, since the

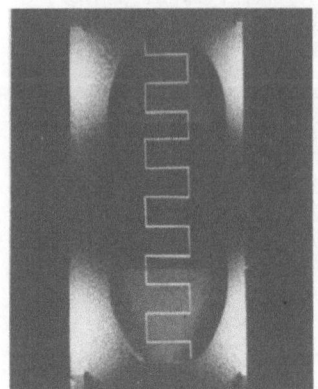 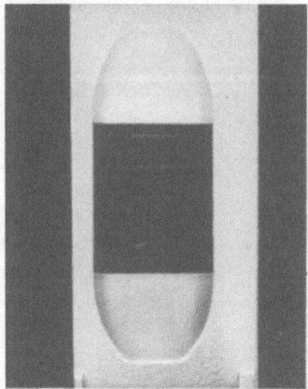

Fig. 12   Photograph demonstrating the effectiveness of a fused plate, with low N.A. and e.m.a., in high ambient lighting conditions: (a) Model consisting of three plates with a simulated phosphor and trace on each. The central plate is a fused plate, N.A. = 0.45, with e.m.a. The two outer plates are plain glass. (b) The system, as shown in a), with ambient lighting. The trace on the plain glass screens is barely visible whilst the central plate still shows a trace of high contrast. (Courtesy of Barr and Stroud Ltd).

image appears on the outside surface, parallax is eliminated and accurate measurements can be made on the trace; a typical CRT is shown in Fig. 13.

Correction of Field Curvature and Distortion

Further direct-vision applications of fused plates are to be found in the correction of field curvature,[1] in which one face of the plate is formed to correspond to the curved field, the other face being flat. This is particularly useful in the opto-electronic field, where field curvature is difficult to correct in the electron optics and where the defects which occur in present-day fused plates are not significant, when compared to defects in the phosphor. Some success has also been achieved in the correction of distortion in the image by imposing an equal and opposite distortion on the fibre distribution in the fused plate. This is done by fusing the fibres together under conditions of non-uniform temperature and pressure. These components are not used in optical systems since the relatively low resolution of the fibre optics degrades the final image significantly. In addition the technique also places a surface at the focal plane which gives further degradation of the image due to defects and dust.

Fig. 13 Photograph of a *CRT*'s fitted with fibre optics faceplates. The tube on the left has a 125 mm × 125 mm faceplate. (Courtesy of M. O. Valve Co. Ltd)

Image Inversion

Some modern image-intensifier systems are fairly short and the length of optical system required to erect the final image can be much greater than the rest of the system. A more elegant approach to this problem is to use a fused plate in which the output faces are rotated 180° with respect to each other; this rotation can be achieved in a length only slightly greater than the plate diameter. To do this the plate is held at its softening temperature and twisted through 180°; during this process the fibres are increased in length in proportion to their distance from the centre of the plate, Fig. 14, and become 'waisted', which lowers the effective N.A. If it is assumed that the developed fibre is of the form of two equal conical frustra, joined at their smaller diameters, one can show that the ratio of the frustrum diameters is given by:

$$\frac{d_2}{d_1} = \left(4\frac{l_i}{l_f} - \frac{3}{4}\right)^{\frac{1}{2}} - \frac{1}{2} \tag{4}$$

where d_1 and d_2 are the frustrum diameters and l_i and l_f are the lengths of the original and final fibre. It can also be shown that:

$$\frac{l_f}{l_i} = \left(4\pi^2\frac{R^2}{l_i} + 1\right)^{\frac{1}{2}} \tag{5}$$

where R is the distance of the fibre from the centre of the plate. It was shown earlier that the effective N.A. of a tapered fibre was given by:

$$(\text{N.A.})_{eff} = \frac{d_2}{d_1}(\text{N.A.}) \tag{6}$$

By combining Eqs. (4), (5) and (6) the curve shown in Fig. 15 is obtained, which shows the variation of the ratio of effective N.A. to theoretical N.A. with radial position of the fibre, where this is measured

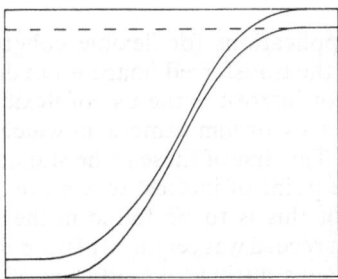

Fig. 14   Side view of a fibre optics image inverter showing the increase in length, and consequent 'waisting', of a typical fibre. The original fibre position is shown as a dotted line.

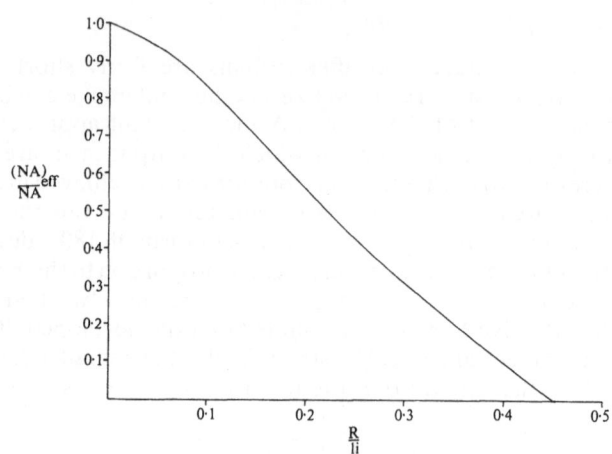

Fig. 15 The variation of the ratio of effective N.A. to theoretical N.A., for a fibre in an image inverter, plotted against radial distance from the centre of the bundle (R/l_i), where l_i is the length of the bundle.

in terms of units of the plate thickness (l_i). This curve shows that the effective N.A. will be zero at a radial distance of about $0.45\,l_i$, at which point the fibre is double its original length and takes the form of two complete cones joined at their apex. In practice, this will not happen since the fibre shape is not truly conical, but the above relation gives a useful guide to the performance of these components.

III. INDIRECT-VISION APPLICATIONS – FLEXIBLE COHERENT BUNDLES

In this section, applications for flexible coherent bundles will be considered in which the transferred image is not directly viewed by the eye. The main area of interest is the use of flexible coherent bundles in conjunction with a TV or film camera, in which there are two main types of application. The first of these is the standard one of removing the camera from the point of interest to a more convenient position. A typical example of this is to be found in the space research programme, where a film record was required of the behaviour of the liquid oxygen in the fuel tanks during take-off. Any violent motion in the tank can lead to an interruption in the supply of liquid oxygen to the motor, causing premature extinction. To simplify the jettison of the cameras, these were mounted on the outside of the space vehicle and the image of the interior of the fuel tanks was transferred to these cameras by means of a flexible coherent bundle.

The second type of application arises from the need to stabilize the camera position in certain environments, e.g. on board aircraft, particularly with TV cameras. If a flexible coherent bundle is connected between the camera and the main objective lens, then only the lens need be stabilized.

IV. INDIRECT VISION APPLICATIONS – SOLID COHERENT BUNDLES

The main application of solid coherent bundles in this area is as fused plates forming the front screen of an opto-electronic device. The prime function of this fused plate is to act as a coupling device to the next system which may be photographic or opto-electronic in nature.

A. Contact Photography

The simplest example of this is the photographic recording of CRT traces in which the fused plate couples the CRT output directly on to the photographic film. Contact photography is not possible with an ordinary CRT since the front screen is a relatively thick glass plate and the phosphor emits light in a Lambertian fashion. Because of this, the photographic record of a trace taken by contact printing from the front surface of a plain glass screen would be a line many times the thickness of the trace, with a Gaussian distribution. Conventionally, traces are recorded by a camera with a high-aperture lens to collect as much light as possible from the trace. However, even the best-quality lens can only collect a small fraction of the available light, whereas a fused plate with an N.A. of unity collects virtually all the available light. Practical measurements have shown that such a plate will collect 30 to 50 times as much light as an oscilloscope camera.[2] Thus events can be recorded by means of a fused plate which are much too fast to be photographed conventionally.

Obviously, to maintain a high resolution, the photographic film must be in close contact with the fused plate. This can be a disadvantage in certain applications, typified by line-scan equipment. In this, a picture is built up in the form of a raster, which is generated by combining the linear motion of a modulated spot on the CRT with a perpendicular motion of the film. If close contact of the film is maintained there is a danger of the plate scratching the photographic emulsion. To overcome this the fused plate is designed to have an N.A. of less than unity, normally around 0.5 to 0.7, which means that the output cone of light from each fibre has a semi-angle of 30 – 45°. Because of this, the film can be placed 50 μm to 100 μm away from the plate surface, with negligible loss in resolution, thus eliminating the possibility of scratching the emulsion and still giving a much higher collection efficiency than a camera. Such plates will, of course, have to be provided with e.m.a.

Image-Intensifier Coupling

The high efficiency of a fused plate for the collection of light from a phosphor is put to further use in the coupling of image-intensifier stages.[3] An image intensifier is a vacuum device in which an optical image is formed on a thin screen of photo-sensitive material deposited on one face of the envelope. This material has a low work function, and the density of electrons emitted from any point on the screen is proportional to the intensity of light at that point. Thus an electronic image is created within the tube which is then focussed on to a phosphor screen on the opposite face. The electrons are accelerated during transit through the tube so that the light intensity emitted from the phosphor is higher than that in the original image; the device is illustrated in Fig. 16.

Typically such a device yields an intensity gain of about × 30, and these can be cascaded to give higher gains. Previously the coupling of these was achieved by using either lenses or thin mica membranes to separate the stages. The former gives poor collection of light but high resolution, whilst the latter gives very good collection of light but an inferior resolution, owing to the finite separation between the phosphor and the subsequent photo-cathode. The use of fibre optics gives

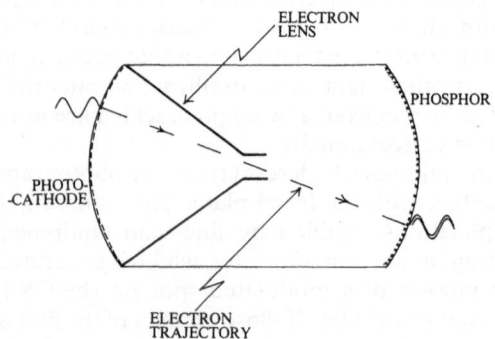

Fig. 16   Diagrammatic representation of the action of an image intensifier tube.

adequate resolution with a high collection efficiency. A cascaded system using fibre-optics coupling is shown diagrammatically in Fig. 17, where the coupling is provided by two fused plates, one forming the output face of one tube and the other the input face of the next tube.

In theory, a single fused plate could be used to couple these stages, but this is not done in practice since the process for deposition of the photo-cathode is not reliable, and the production of two satisfactory photo-cathodes successively cannot be guaranteed. The stages are therefore manufactured as separate units and acceptable units coupled together. The stages are coupled through a thin film of oil between the fused plates to eliminate losses due to Fresnel reflection. The internal surfaces of the fused plates are curved to allow for the field curvature in the electron optics system. Since a high collection efficiency is required, these plates are designed with a high N.A. The internal face of the plates can have a relatively large curvature so that the end faces of the fibres towards the periphery of the plates are not normal to the fibre axis. Therefore, in order to maintain a high collection efficiency from these fibres, the N.A. is normally in excess of unity, typically 1.1 1.2. (Section 2.II.c)

The photo-cathode is formed from a mixture of alkali metals which creates materials problems, since these chemically reduce some of the oxides used in optical glass manufacture, particularly lead oxide. The effect of this is to create an opaque metallic film on the screen which also reduces the photo-electronic conversion efficiency, and so the glasses used for these plates are lanthanum borates which are lead-free.

The resolution of a number of fused plates used in series will obviously be less than that of a single plate of the same thickness, since the output from one fibre will in general enter a number of fibres in the following plate. It can be shown that the resolution of a number of similar fused plates (R_n) used in parallel is given by:

$$R_n = \frac{R_0}{\sqrt{n}}$$

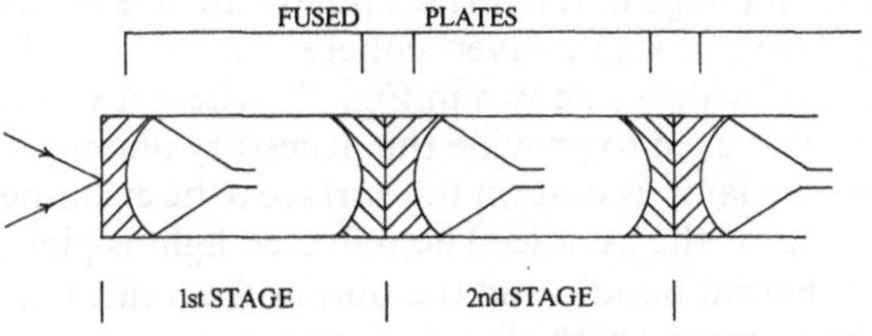

Fig. 17 The cascading of image intensifiers, using fused plates.

where R_0 is the resolution of a single plate and n is the number of plates. Thus, a three-stage image intensifier will suffer a loss in resolution by a factor of approximately two owing to the fibre optics, compared with a single-stage tube. For this reason, a very high resolution is required from these plates and the individual fibre sizes are $5-7\,\mu$m giving resolutions of 80–100 line-pair/mm. A three-stage image-intensifier system has a gain of 3×10^4 which means that scenes can be rendered visible to the eye through this system, where the illumination is provided by starlight only.[4] As a result, this system has an important military application as a night sight. A scene as viewed through such a device is shown in Fig. 18.

Increasing use is being made of fused plates as the front face of TV camera tubes, so that an image-intensifier system incorporating fibre optics can be coupled directly into the tube. This greatly increases the capability of TV cameras to cover outside events and to survey secure areas at night, e.g. dockyards and warehouses.

V. MISCELLANEOUS APPLICATIONS – COHERENT BUNDLES

There are a number of applications for coherent bundles, in which the fibre assembly is not the standard one as described above. These are akin to the beam-splitting and shape-changing applications for non-coherent bundles but, because of the difficulties involved in the manufacture of this type of bundle, the applications are not numerous and this area can be regarded as being in its infancy. A few representative applications are outlined below.

A. Surface Scanning

This application is typified by the inspection of the internal surfaces of heat exchangers designed for use with nuclear power, which are built up from rectangular ducts approximately 50 mm by 2 mm internally. It is obviously impossible to inspect the internal surfaces of these ducts by conventional means and the most efficient method involves fibre optics. In this, an image of the surface is built up, line by line, on photographic film using a single-layer coherent bundle, a diagrammatic representation of which is shown in Fig. 19, where it will be noted that fibre optics is also used to provide the necessary illumination.[5] In use, the light from the lamp is cast on the surface to be examined by the 45° mirror at the tip of the bundle. The reflected light is picked up by the single-layer coherent bundle and the output from this bundle is focussed on the film by means of the lens to form a line image. The resolution

Fig. 18 Photograph of the output from a three-stage image intensifier, coupled by fused plates. The illumination is equivalent to a moonless, starlit sky. The pincushion distortion of the image is caused by the electron optics and is removed in the viewed image by introducing equal and opposite distortion into the viewing eyepiece. (Crown copyright reserved. Reproduced by permission of the Controller, H. M. Stationary Office.)

of this line image is less than the theoretical value, since the surface is not in contact with the input face of the coherent bundle. However, by using a relatively low N.A. (around 0.5), the discrepancy is small for the fibre size used, which is 125 μm. The image is then built up by moving the tip of the bundle over the surface and, at the same time, moving the film in synchronism. If the film speed equals the scanning speed, then a record will be obtained at a magnification of unity, if the lens also

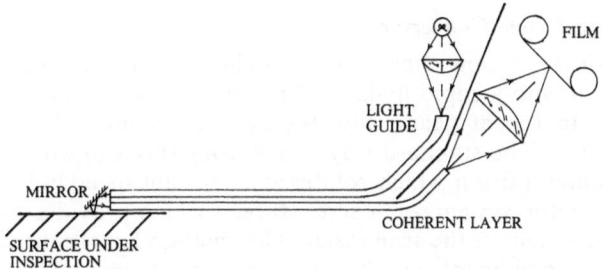

Fig. 19 The optical system for a proximity scanning device.

Fig. 20   A typical photographic record, obtained by the device illustrated in Fig. 18.

operates at this magnification. Obviously, by altering the lens magnification and the ratio of scanning speed to film speed in the same proportion, one can obtain any desired magnification. A typical record from such an instrument is shown in Fig. 20.

B. Line-to-Circle Converter

In optical scanning there has always been a problem in achieving a linear scan with a rapid fly-back. This problem has been given some attention in recent years, with the upsurge of interest in facsimile transmission. The standard way of achieving this is to wind the document around a drum which rotates at a constant speed in front of the photo-detector system, on a screw thread which provides the orthogonal movement of the scan raster. This method has the disadvantage that not all documents can be wound round a drum and, in any case, this requires a degree of training and skill on the part of the operator which cannot be guaranteed.

By the use of a coherent bundle, a flat scan can be achieved which requires no skill from the operator and permits transmission from bulky documents, e.g. books. The type of bundle is shown diagrammatically in Fig. 21 and consists of a single-layer coherent bundle whose width equals the maximum width of document to be scanned. One end of the bundle is mounted on the bed of the equipment and is straight, whilst the other is formed into a circle so that the first fibre is very close to the last fibre in the layer. The line image picked up by the straight end is therefore formed into a circle which can be scanned linearly by rotating a detection system over this end with the same radius and centre of rotation. In practice, this scanning is done by a single fibre, as shown in the diagram, so that the photo-detector can be stationary on the axis of rotation. The illumination for such a system is provided by a light guide with a circular input and rectangular output, with one side of the rectangle equal in length to the input face of the coherent bundle. The rectangular output is positioned parallel and close to the coherent-bundle input (Fig. 22). In some designs a reasonable distance is left between the first and last fibre in the circle, so that a small light source can be positioned between these to provide a synchronization pulse once every revolution.

One disadvantage of the fibre-optics system is that a broken fibre appears as a black line in the facsimile. Since a typical bundle might have 5000 fibres, it is inevitable that a few of these will be broken. One way of overcoming this is to arrange that the fibre size is half that required by the resolution of the system; then one can scan using an aperture equal to two fibre diameters and arrange the detector to register black only when the signal level is below 50%. This arrangement is adequate when half-tones are not required in the facsimile, but if half-tones are required, then the scanning is done with an aperture equal to the fibre size and the resulting output is processed by logic circuitry which only allows black to be registered if two adjacent fibres are not transmitting. Both of these techniques reduce the specification

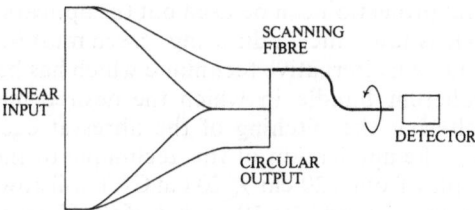

Fig. 21 Diagram of a coherent circle-to-line converter, showing the use of a fibre bundle to perform the scan.

Fig. 22   Photograph of a coherent circle-to line converter capable of scanning paper widths up to 25 cm. (Courtesy of Rank Precision Industries Ltd).

for breakages from zero breakages, which is almost impossible to achieve, to a requirement that no broken fibres are to be adjacent, which is an acceptable condition.

C. Large-Scale Displays

In an age where effective communication is vital there has been a great deal of development of means for displaying information. Unquestionably, the CRT is the best means yet available for displaying large amounts of information; however, this device is restricted in size and cannot be conveniently used where the information display has to be large. Optical projection can be used but the optical system is bulky and light output is low, since a diffusing screen must be placed at the final image plane. An alternative technique which has been developed is to use a coherent bundle in which the desired magnification is achieved by altering the pitching of the fibres at each end. As an example of this, the application of this technique to the provision of a 2 m × 2m display from a 20 cm × 20 cm CRT will now be described. The magnification required is × 10 so that, if the fibres are touching at the input face, they must be pitched apart 10 fibre diameters at the output, and the fibre size is therefore chosen so that the eye cannot

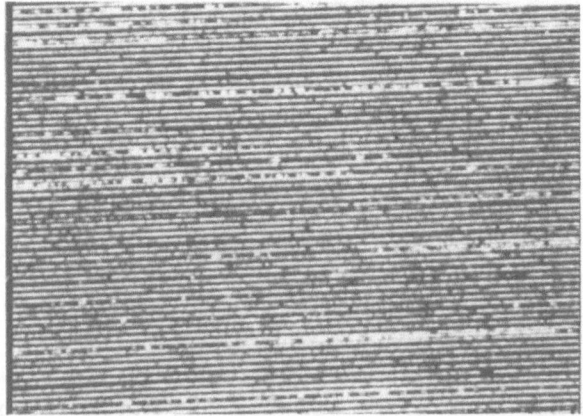

Fig. 23 Photograph of the consolidated coherent fibre layer used in the manufacture of the magnifier. The fibre size is 100 μ.

detect this pitching at the intended viewing distance. In this example, the display was to be viewed at a distance of several metres and a 1 mm final pitching was chosen, which required a fibre diameter of 0.1 mm.

The manufacture of the bundle is carried out in the following manner. A single coherent layer of 0.1 mm fibres, which is 20 cm wide, is wound on a drum with a diameter of 2m. The fibres are bonded together on the drum by spraying a thin plastic film on the layer, which can then be sliced and removed from the drum as a sheet which is approximately 6 m long (Fig. 23). The plastic is removed from a diagonal strip about 1 cm wide, on this layer, the length of the strip being 2 m. The position of this is central on the layer, as is shown in Fig. 24. If the fibres in this strip can be bent so that they are at right angles to it, then the pitching of the fibres will be 1 mm. This is illustrated in Fig. 25 and is achieved by laying the sheet down flat on a smooth surface and sliding the two bonded portions of the sheet in a direction parallel to the diagonal strip. This motion is continued until the fibres cross the centre line of the diagonal strip normally. When this is done, the pitching of the fibres will be 1.0 mm on average, although there will be some spread around

Fig. 24 Diagram showing the diagonal strip of fibre left unbonded in the coherent layer used to make a magnifying coherent bundle.

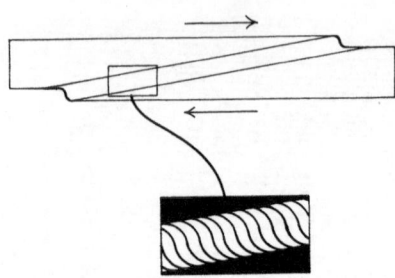

Fig. 25   Diagram showing the manipulation of the layer, shown in Fig. 24, to create the required fibre pitch.

this figure. The ensure accurate pitching, a jig with slots pitched 1 mm apart is moved up to locate the fibres precisely. The fibres in this area are then set in resin and the sheet cut in two along the centre line of the diagonal strip. This fibre layer will achieve the desired magnification in one direction. In the orthogonal direction this is achieved by stacking a number of such layers so that the layers are touching at the input end, but are pitched 1.0 mm apart at the output end by means of spacers. The finished bundle has an input face measuring 20 cm × 20 cm and an output face measuring 2 m × 2 m. The CRT can be directly coupled into the input face if a fibre-optics fused plate has been fitted.

   This manufacturing technique, in addition to being an elegant solution to the problem of pitching the fibres, gives a display unit which has a small depth. In the above example, the depth of the unit would be less than 1.0 m as illustrated in Fig. 26.

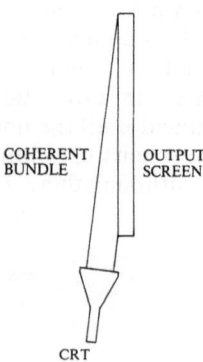

COHERENT
BUNDLE

OUTPUT
SCREEN

CRT

Fig. 26   Finished coherent bundle magnifier coupled to a *CRT*.

D. Coding and Decoding

In certain areas of security it would be an advantage to be able to code and decode visual information. Fibre optics offers an immediate answer in the form of a bundle in which the position of fibres at one end bears no simple relation to their position at the other end. Whilst one such bundle can be easily manufactured it is difficult to make the two identical bundles which would be required for the coding and decoding operation.

A modification to this technique has been developed by the American Optical Co. Ltd, in which the coding is carried out in one dimension only. In such a bundle, coherent layers of fibre are used which give image transfer in one dimension and are built up into a bundle in which the layers are assembled conventionally at one end but are staggered with respect to one another at the other end. This staggering is carried out in a predetermined manner and can be repeated in any number of bundles. An image which is transferred through such a bundle appears as a series of disjointed points. This is illustrated in Fig. 27 for a six-layer bundle, although a much larger number of layers would be used in practice to give effective coding. It would be possible, although very tedious, to decode the image without the decoding bundle, but the coding can be made virtually secret by a second coding operation using a different distribution of stagger.

VI. ENHANCEMENT OF THE TRANSFERRED IMAGE

Since the image transferred through a coherent bundle is coincident with the output face of that bundle, the fibre structure is apparent if the image is viewed under high manification. The optimum magnification is where the fibre structure is just noticeable and a higher magnification will make this structure more apparent, without increasing the amount

Fig. 27   Diagram showing the coding of the word 'FILE' using a rudimentary 6-layer bundle.

of detectable detail in the image. Even at the optimum magnification, the fibre structure can be distracting under critical viewing conditions and a number of ways of eliminating this have been devised.

A. Spatial Filtering

Since the transferred image contains no frequencies above that determined by the fibre pitching, it should be possible to eliminate the fibre structure from the image by using a spatial filter which cuts out frequencies above a critical frequency which would be set just below that of the fibre structure.[6] Ideally this filter should possess an infinitely sharp cut-off at this frequency; however, this is not achievable in practice and some loss of contrast in the image occurs at lower frequencies. In fact, a similar result can be achieved by de-focussing the system slightly.

B. Time-shared multiplexing

A better approach is to use the principle of dynamic scanning, which was outlined in Chapter 6; in this the bundle faces are moved relative to the image in synchronism. If the sampling interval allows each information point to be crossed by a large number of fibres, then the resultant response will be the average response of a finite area of the bundle face. The larger this area is the more nearly constant will be the response of the bundle at each information point, and the less will be the effect of the fibre structure. For direct viewing the sampling interval is determined by the frequency response of the human eye. An additional benefit is that the overall resolution of the bundle will be almost doubled, as was discussed in Chapter 6.

To give continuous multiplexing a cyclic motion has to be given to the scan and one means of achieving this is illustrated in Fig. 28 where the image is moved relative to the bundle of means of a rotating thin prism, which imparts a circular motion to the image. This is removed at the output by means of a similar prism contra-rotating at the same speed. Multiplexing is normally carried out over about 10 structural units since it is found that greater sampling areas give little or no gain

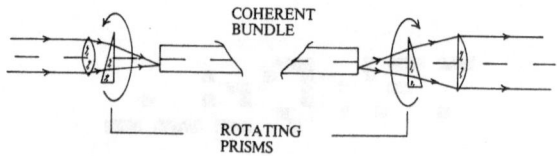

Fig. 28   Time-shared multiplexing, using thin prisms.

in quality. This means that with a flexible coherent bundle, made from 60 μm square multiple fibres, an area of incomplete multiplexing occurs up to 0.6 mm from the edges of the bundle. It is therefore not normally worth multiplexing bundles if their cross-section is less than 4 mm square, since the area of incomplete multiplexing would then be inconveniently large.

It is obvious that the effect of discontinuous variations in structure, e.g. fibre breakages, will also be reduced by multiplexing in the ratio of the total number of fibres in the sampling area to the number of breakages in that area. This ratio will be about 100:1 in most commercially-available bundles, which have breakages occurring at the rate of about one breakage in every hundred fibres. This method gives very good results, but the necessity of providing some form of rotatrional motion at the objective end means that the instrument tends to be bulky. An elegant alternative is the following.

C. Wavelength Multiplexing

In this the transference of image points is shared among a large number of fibres by dispersive means.[7] This is illustrated in Fig. 29, where a direct-vision prism is used to split each image point into a line spectrum, and is then recombined into a single point at the output by passing the emergent light through a similar prism. There are no moving parts to this system and the instrument design is virtually unaltered. However, the multiplexing only takes place in one dimension, so that structural information at right angles to the direction of dispersion is unaltered. In spite of this a very acceptable image is formed, the main criticism of which is that broken fibres show up as slight 'smears' of colour.

In both cases of multiplexing, the edges of the transferred image are not fully multiplexed since the sampling area, at the edge, includes regions where no fibres are present. This is perhaps more serious in the case of wavelength multiplexing, since the edges of the image at right angles to the dispersion are coloured, owing to the non-transference of the extreme portions of the line spectra at these edges. This area of incomplete multiplexing is the width of the scan, in the first

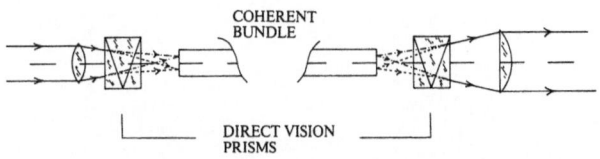

Fig. 29 Wavelength multiplexing, using direct vision prisms. Upper broken lines show the path of red light, the lower, that of the blue light, from the image.

case, of the line spectra in the second and is blocked by a mask in the eyepiece. An increased sampling area, and, therefore, a better multiplexing will result in a larger area of incomplete multiplexing.

REFERENCES

1. N. S. Kapany and R. E. Hopkins, *J. Opt. Soc. Am.* **47**, 1109 (1957)
2. L. S. Allard, *Ind. Electronics* **2**, 273 (1964)
3. N. S., *Proc. 2nd. Image Intensifier Symposium*, Fort Belvoir, Va. (1961) p. 143.
4. P. J. Dolon and W. F. Niklas *Proc. 2nd Image Intensifier Symposium*, Fort Belvoir, Va. (1961), p. 93.
5. D. F. Capellaro, *Light and Heat Sensing: AGARDograph 71* (H. J. Merrill, Ed.), Pergamon Press, Oxford (1963), p. 311
6. N. S. Kapany, *Sci. Am.* **203**, (No. 5), 72 (1960)
7. C. J. Koester, *J. Opt. Soc. Am.* **58**, 63 (1968)

CHAPTER 9

Waveguide Properties of Optical Fibres

In previous chapters the properties of the optical fibre have been described using the results of geometric optics. However, these can also be described in the language of physical optics, and, in certain instances, this approach is the only valid one. In this respect, the optical fibre is no different from any other optical system.

The above statements are particularly true when the fibre diameter is of similar magnitude to the wavelength of the light being transmitted. Under these circumstances the behaviour of the light beam must be described in terms of electromagnetic theory and the fibre must be considered as a waveguide. Some interest has been shown in this aspect of fibre optics by workers in the communications field, owing to the high information capacity of a light beam, and this application may prove to be important in the future.

I. THE WAVEGUIDE AS AN OPTICAL SYSTEM

The normal approach to the description of a guided electromagnetic wave is to solve Maxwell's field equations for the boundary conditions imposed by the waveguide.[1] However, it is instructive to consider this situation from an optical viewpoint and regard the guided wave as the interference pattern set up by an electromagnetic wave which is reflected at the walls of the guide.

To simplify the discussion, we will consider a plane-polarized wave reflected by two infinite metallic sheets, which are parallel to each other and spaced a distance, a, apart. It will be assumed that the direction of propagation is the Z-direction and that one of the sheets coincides with the Y Z plane. A plane-polarized wave travelling with velocity u, in a direction represented by a unit vector $\mathbf{n}$, can be described in terms of its electric vector by the expression:

$$\mathbf{E} = \mathbf{E}_0 \exp \left\{ i\omega \left(t - \frac{\mathbf{r}.\mathbf{n}}{u} \right) \right\}$$

where $\mathbf{r}$ is the position vector, ω is the angular frequency, and $\mathbf{E}_0$ is the amplitude.

Since the wave is plane:

$$\mathbf{n}.\mathbf{E} = 0$$

The wave incident into the guide, as shown in Fig. 1 with an angle of incidence ϕ, can be represented as:

$$\mathbf{E}_1 = (E_0 \sin \phi \mathbf{i} - E_0 \cos \phi \mathbf{k}) \exp. \left\{ i\omega \left(t - \frac{x \cos \phi}{u} - \frac{z \sin \phi}{u} \right) \right\} \quad (1)$$

where $\mathbf{i}$ and $\mathbf{k}$ are unit vectors in the x and z directions respectively. Similarly, the reflected wave can be represented by:

$$\mathbf{E}_2 = (E_0 \sin \phi \mathbf{i} + E_0 \cos \phi \, \mathbf{k}) \exp \left\{ i\omega \left(t + \frac{x \cos \phi}{u} - \frac{z \sin \phi}{u} \right) \right\} \quad (2)$$

The field distribution resulting from the interference of these two will be given by:

$$\mathbf{E}_T = \mathbf{E}_1 + \mathbf{E}_2$$

$$= \left[E_0 \sin \phi \mathbf{i} \left\{ \exp \left(\frac{i\omega \cos \phi}{u} x \right) + \exp \left(-\frac{i\omega \cos \phi}{u} x \right) \right\} + \right.$$

$$\left. E_0 \cos \phi \mathbf{k} \left\{ \exp \left(\frac{i\omega \cos \phi}{u} x \right) - \exp \left(-\frac{i\omega \cos \phi}{u} x \right) \right\} \right] \times$$

$$\exp \left\{ i\omega \left(t - \frac{z \sin \phi}{u} \right) \right\}$$

$$= \left[2 E_0 \sin \phi \cos \left(\frac{\omega \cos \phi}{u} x \right) \mathbf{i} + 2 i E_0 \cos \phi \sin \left(\frac{\omega \cos \phi}{u} x \right) \mathbf{k} \right] \times$$

$$\exp \left\{ i\omega \left(t - \frac{z \sin \phi}{u} \right) \right\} \quad (3)$$

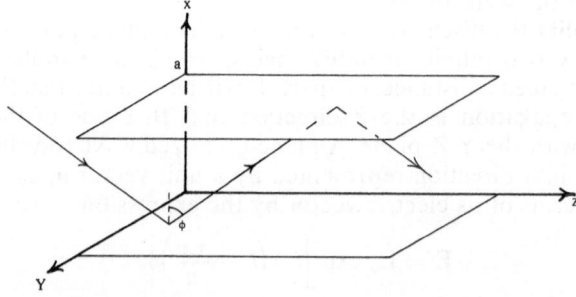

Fig. 1   Illustration of the incidence of an electro-magnetic wave into a parallel plate wave guide, with plate separation, a.

where the exponentials have been replaced by their trigonometric equivalents. Thus:

$$\mathbf{E}_T = 2\,E_0 \sin\phi \cos\left(\frac{\omega\cos\phi}{u}x\right)\exp\left\{i\omega\left(t - \frac{z\sin\phi}{u}\right)\right\}\mathbf{i} \;+\;$$
$$2\,iE_0 \cos\phi \sin\left(\frac{\omega\cos\phi}{u}x\right)\exp\left\{i\omega\left(t - \frac{z\sin\phi}{u}\right)\right\}\mathbf{k} \tag{4}$$

It will be seen from Eq. (4) that $\mathbf{E}_T$ has an x and z component given by:

$$E_x = 2\,E_0 \sin\phi \cos\left(\frac{\omega\cos\phi}{u}x\right)\exp\left\{i\omega\left(t - \frac{z\sin\phi}{u}\right)\right\}$$
$$E_z = 2\,iE_0 \cos\phi \sin\left(\frac{\omega\cos\phi}{u}x\right)\exp\left\{i\omega\left(t - \frac{z\sin\phi}{u}\right)\right\} \tag{5}$$

Since the amplitude is arbitrary one can, without loss of generality, divide both equations in (5) throughout by $2\,iE_0 \cos\phi$ giving:

$$E_x = -i\tan\phi\,\cos\left(\frac{\omega\cos\phi}{u}x\right)\exp\left\{i\omega\left(t - \frac{z\sin\phi}{u}\right)\right\}$$
$$E_z = \qquad\quad \sin\left(\frac{\omega\cos\phi}{u}x\right)\exp\left\{i\omega\left(t - \frac{z\sin\phi}{u}\right)\right\} \tag{6}$$

The boundary conditions require that $E_z = 0$ when $x = 0$ and $x = a$. The first of these is already fulfilled by the choice of the initial wave function and the second requires that:

$$\frac{\omega\cos\phi}{u}a = m\pi$$
$$\text{i.e.}\quad \frac{\omega\cos\phi}{u} = \frac{m\pi}{a} \tag{7}$$

where m is an integer.

Substituting Eq. (7) in Eq. (6) and letting $\dfrac{\omega\sin\phi}{u} = \beta$ gives:

$$E_x = -i\tan\phi\,\cos\left(\frac{m\pi}{a}x\right)\exp\left\{i(\omega t - \beta z)\right\}$$

and,

$$E_z = \sin\left(\frac{m\pi}{a}x\right)\exp\left\{i(\omega t - \beta z)\right\} \tag{8}$$

where, since $\sin\phi = \dfrac{u\beta}{\omega}$, and $\cos\phi = \dfrac{u}{\omega}\cdot\dfrac{m\pi}{a}$, E_x can be written as:

$$E_x = -\frac{i\beta a}{m\pi}\cos\left(\frac{m\pi}{a}x\right)\exp\left\{i(\omega t - \beta z)\right\} \tag{9}$$

Eq. (8) and Eq. (9) are identical with the field distributions obtained by the solution of Maxwell's equations for a transverse magnetic wave propagated along the z axis with a propagation constant β, where the defining parameters of the distribution are related by:

$$\frac{\omega^2}{u^2} = \left(\frac{m\pi}{a}\right)^2 + \beta^2 \tag{10}$$

The planes of constant phase for the wave will be given by the expression:

$$\omega t - \beta z = \text{constant}$$

Therefore the phase velocity, v', will be given by:

$$v' = \frac{dz}{dt} = \frac{\omega}{\beta}$$

Which gives, from the definition of β:

$$v' = \frac{u}{\sin \phi} \tag{11}$$

The planes of constant group amplitude are defined by the expression:

$$t\,d\omega - z\,d\beta = \text{constant}$$

so that the group velocity, u', is given by:

$$u' = \frac{dz}{dt} = \frac{d\omega}{d\beta}$$

$$= \frac{u^2}{v'} \qquad \text{from (10)}$$

$$= u \sin \phi \tag{12}$$

The values for group and phase velocity are those which would have been deduced from an inspection of Fig. 1, and the relation between these is illustrated in Fig. 2.

As $\phi \longrightarrow \pi/2$, i.e. grazing incidence, then $u' \longrightarrow v' \longrightarrow u$, and the wave behaviour tends to that of a free, i.e. unguided, wave. As $\phi \longrightarrow 0$, i.e. normal incidence, $u' \longrightarrow 0$ and $v' \longrightarrow \infty$, and the wave is reflected back and forth between the plates with no forward motion.

For unattenuated propagation β must be real; therefore we can write, from Eq. (10):

$$\frac{\omega^2}{u^2} > \left(\frac{m\pi}{a}\right)^2$$

The critical 'cut-off' frequency will be given by:

$$\omega_0 = \frac{m\pi}{a}u$$

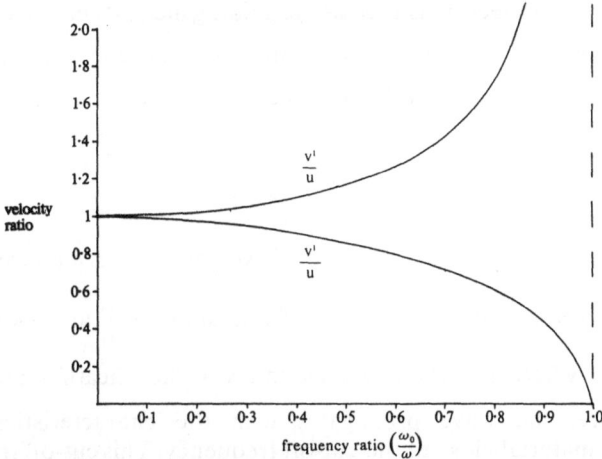

Fig. 2 Variation of the phase velocity ratio (v'/u) and the group velocity ratio (u'/u), where u is the unbounded velocity, for a guided electro-magnetic wave, plotted as a function of the ratio of cut-off frequency to wave frequency (ω_0/ω).

where, for frequencies less than this, unattenuated propagation is impossible. We can re-write β and ϕ in terms of this critical frequency:

$$\beta = \frac{\omega}{u}\left(1 - \left(\frac{\omega_0}{\omega}\right)^2\right)^{\frac{1}{2}}$$

and

$$\sin \phi = \left(1 - \left(\frac{\omega_0}{\omega}\right)^2\right)^{\frac{1}{2}}$$

where we see that at the critical frequency:

$$\sin \phi = 0 \text{ and } \beta = 0$$

In this case, the group velocity is zero and a stationary wave pattern is produced.

It can be seen therefore that, for any given frequency above cut-off, there are specific angles of incidence which will enable a freely propagating pattern to be set up, determined by the permitted values of the integer m. As the frequency increases, the number of possible modes increases and, for frequencies much higher than the critical value, virtually all possible incidence angles can exist. Under these conditions, the system can be adequately described using the geometric optics.

If internal reflection is used as the guiding means then the velocity of
the unbounded wave in the medium is $\frac{c}{n_1}$, where n_1 is the refractive
index of the 'core' material and the phase velocity of the pattern is
given by:

$$v' = \frac{c}{n_1 \sin \phi}$$

This approaches the 'unbounded' velocity, $\frac{c}{n_1}$, as ϕ approaches $\frac{\pi}{2}$,

i.e. far from cut-off. On the other hand, $\sin \phi = \frac{n_2}{n_1}$ at cut-off so that

$v' = \frac{c}{n_2}$, where n_2 is the refractive index of the 'sheath' material.

Therefore, the wave propagates with the characteristics of the
'sheath' material close to the cut-off frequency. This cut-off frequency
is not so well-defined as is the case for metallic reflection, since unattenuated propagation is still possible for angles less than the critical
angle for internal reflection. However, these rays will not be bound to
the core and are not regarded as acceptable modes within the context
of this discussion.

II. WAVEGUIDE MODES IN AN OPTICAL FIBRE

The treatment of the possible modes within an optical fibre is made
extremely complex by the fact that the field distribution is not confined within the core and separate sets of solutions of the field equations must be formulated for the core and sheath respectively. In this
case, the boundary conditions are that these solutions must be compatible at the boundary. In addition, the boundary conditions require
that a transverse magnetic and a transverse electric wave be superimposed to give an acceptable solution. This means that, in general,
the resultant wave is hybrid, i.e. there is a non-zero component of both
E and H in the direction of propagation.

The complete theory of the propagation of waves within a dielectric
cylinder was set out by Hondros and Debye[2] and this has been applied
to the case of an optical fibre by Snitzer[3] and Kapany.[4] However, the
optical fibres which are of interest as optical waveguides have a very
small refractive index difference between the core and sheath, to
restrict the number of bound modes, and this fact can be used to obtain
approximate solutions of the wave equation. The treatment presented
here is due to Gloge[5] and the approximate solutions obtained are
exact at cut-off and are in error by only a few percent for conditions
far from cut-off.

The system to be considered consists of an optical fibre with core radius a and a core refractive index n_1; the sheath is supposed to be infinitely thick with a refractive index of n_2. Within this system, the propagation constant, β, of a bound mode will lie within the range:

$$n_1 k \gg \beta \gg n_2 k$$

where $k \left(= \dfrac{2\pi}{\lambda} \right)$ is the wavenumber in free space. This is identical with the propagation conditions outlined in the previous section. It can be shown that the field distribution within the core can be expressed by a Bessel function $J\left(\dfrac{ur}{a}\right)$, Fig. 3, where $u = a(k^2 n_1^2 - \beta^2)^{\frac{1}{2}}$.

Similarly, the field distribution within the sheath can be expressed by a modified Hankel function $K\left(\dfrac{wr}{a}\right)$, Fig. 4, where:

$$w = a(\beta^2 - k^2 n_2^2)^{\frac{1}{2}}$$

These functions decrease rapidly to zero as the argument increases and can therefore be used to describe a field bound to the core. The parameters u and w are combined to provide a third parameter v, defined by:

$$v = (u^2 + w^2)^{\frac{1}{2}} = ak(n_1^2 - n_2^2)^{\frac{1}{2}} \tag{13}$$

which can be regarded as a normalized frequency.

These distributions must be continuous at the core–sheath interface and this condition yields characteristic functions $u(v)$ or $w(v)$ which

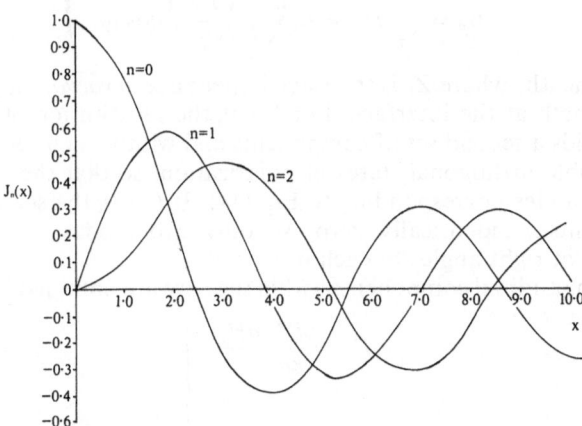

Fig. 3   Variation of the Bessel function $J_n(x)$, for the first three orders, plotted against x.

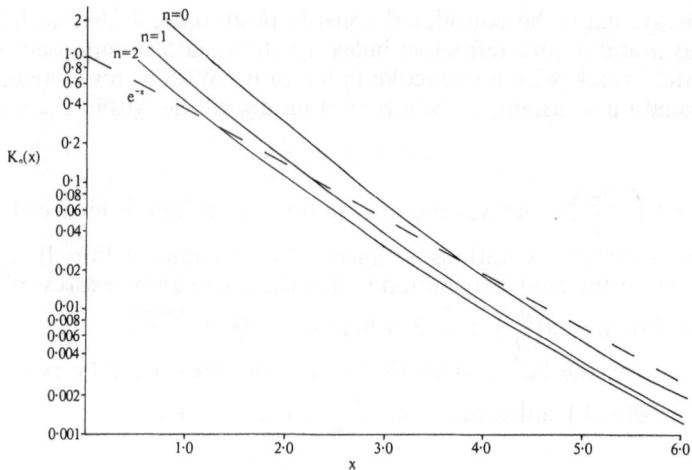

Fig. 4   Variation of the modified Hankel function $K_n(x)$, for the first three orders, plotted against x. The function e^{-x} is also shown as a dotted line.

define the mode. If the fibre axis is coincident with the z co-ordinate axis, transverse field components can be postulated as follows:

$$E_y = \frac{Z_0}{n_1} H_x = E_l \left(\frac{J_l\left(\frac{ur}{a}\right)}{J_l(u)} \right) \cos l\phi$$

for the core, and

$$E_y = \frac{Z_0}{n_2} H_x = E_l \left(\frac{K_l\left(\frac{wr}{a}\right)}{K_l(w)} \right) \cos l\phi$$

(14)

for the sheath, where Z_0 is the wave impedance *in vacuo*, and E_l is the field strength at the interface. For $l > 0$, the substitution of $\sin l\phi$ for $\cos l\phi$ yields a second set of components and within each set there are two possible orthogonal states of polarization, so that there are four possible modes corresponding to Eq. (14). If $l = 0$, the second set of components is identically zero, so only two modes are possible, polarized at right angles to each other.

The longitudinal components within the core are obtained from:

$$E_z = \frac{iZ_0}{kn_1^2} \frac{\partial H_x}{\partial y}$$

and

$$H_z = \frac{ik}{Z_0} \frac{\partial E_y}{\partial x}$$

(15)

which gives:

$$E_z = -\frac{iE_l}{2ka}\left\{ \frac{u}{n_1}\frac{J_{l+1}\left(\frac{ur}{a}\right)}{J_l(u)}\sin(l+1)\phi + \right.$$

$$\left. \frac{u}{n_1}\frac{J_{l-1}\left(\frac{ur}{a}\right)}{J_l(u)}\sin(l-1)\phi \right\} \tag{16}$$

$$H_z = -\frac{iE_l}{2kZ_0a}\left\{ u\frac{J_{l+1}\left(\frac{ur}{a}\right)}{J_l(u)}\cos(l+1)\phi - \right.$$

$$\left. u\frac{J_{l-1}\left(\frac{ur}{a}\right)}{J_l(u)}\cos(l-1)\phi \right\}$$

These components are smaller than the transverse components by a factor of $\frac{u}{ka} = (n_1^2 - \beta^2/k^2)^{\frac{1}{2}} \approx (n_1^2 - n_2^2)^{\frac{1}{2}}$. The derivation of the transverse components from the above equations yields components which are smaller than those postulated by a factor of $\left(\frac{u}{ka}\right)^2$, i.e. approximately $n_1^2 - n_2^2$. Since $(n_1 - n_2) \ll 1$, these can be ignored and it will be assumed that the modes are linearly polarized, with field distributions given by Eq. (14) and Eq. (16). A similar argument applies to the field distribution within the sheath which is given by:

$$E_z = -\frac{iE_l}{2ka}\left\{ \frac{w}{n_2}\frac{K_{l+1}\frac{wr}{a}}{K_l(w)}\sin(l+1)\phi - \frac{w}{n_2}\frac{K_{l-1}\left(\frac{wr}{a}\right)}{K_l(w)}\sin(l-1)\phi \right\} \tag{17}$$

$$H_z = -\frac{iE_l}{2kZ_0a}\left\{ w\frac{K_{l+1}\left(\frac{wr}{a}\right)}{K_l(w)}\cos(l+1)\phi + w\frac{K_{l-1}\left(\frac{wr}{a}\right)}{K_l(w)}\cos(l-1)\phi \right\}$$

The transverse components can be rewritten in cylindrical co-ordinates as:

$$E_\phi = \tfrac{1}{2}E_l\left(\frac{J_l\left(\frac{ur}{a}\right)}{J_l(u)}\right)\left\{\cos(l+1)\phi + \cos(l-1)\phi\right\} \tag{18}$$

$$H_\phi = -\tfrac{1}{2}\frac{E_l}{Z_0}\left(n_1 \cdot \frac{J_l\left(\frac{ur}{a}\right)}{J_l(n)}\right)\left\{\sin(l+1)\phi - \sin(l-1)\phi\right\}$$

for the core and:

$$E_\phi = \tfrac{1}{2}E_I \left(\frac{K_I\left(\frac{wr}{a}\right)}{K_I(w)} \right) \left\{ \cos{(l+1)\phi} + \cos{(l-1)\phi} \right\}$$

$$H_\phi = -\tfrac{1}{2}\frac{E_I}{Z_0}\left(n_2 \frac{K_I\left(\frac{wr}{a}\right)}{K_I(w)} \right) \left\{ \sin{(l+1)\phi} - \sin{(l-1)\phi} \right\}$$

(19)

for the sheath.

The tangential distributions described by Eqs. (16) and (17), (18) and (19) must equate at $r = a$, in order to satisfy the continuity condition at the core – sheath interface. This automatically the case in Eqs. (18) and (19), and Eqs. (16) and (17) yield the following condition for the continuity of electric field:

$$\frac{u}{n_1}\frac{J_{l+1}(u)}{J_l(u)}\sin{(l+1)\phi} + \frac{u}{n_1}\frac{J_{l-1}(u)}{J_l(u)}\sin{(l-1)\phi}$$

$$= \frac{w}{n_2}\frac{K_{l+1}(w)}{K_l(w)}\sin{(l+1)\phi} - \frac{w}{n_2}\frac{K_{l-1}(w)}{K_l(w)}\sin{(l-1)\phi}$$

and a similar expression for the magnetic field. Since $n_1 - n_2 \ll 1$, n_1 and n_2 can be regarded as equal and may be cancelled from both sides, and by using the recurrence relations,

$$2\frac{lJ_l(u)}{u} = J_{l-1}(n) + J_{l+1}(u)$$

and

$$-2\frac{lK_l(w)}{w} = K_{l-1}(n) - K_{l+1}(w)$$

this condition becomes:

$$\left(-u\frac{J_{l-1}(u)}{J_l(u)} + 2l \right)\sin{(l+1)\phi} + u\frac{J_{l-1}(u)}{J_l(u)}\sin{(l-1)\phi}$$

$$= \left(w\frac{K_{l-1}(w)}{K_l(w)} + 2l \right)\sin{(l+1)\phi} - w\frac{K_{l-1}(w)}{K_l(w)}\sin{(l-1)\phi}$$

which is obviously satisfied if:

$$u\frac{J_{l-1}(u)}{J_l(u)} = -w\frac{K_{l-1}(w)}{K_l(w)}$$

(20)

It can be shown that this relation also satisfies the magnetic field continuity condition. Thus Eq. (20) is the characteristic equation for these linearly-polarized modes, with the assumption that $n_1 - n_2 \ll 1$.

The cut-off conditions for the modes are obtained from Eq. (20) by setting $w = 0$, which gives:

$$J_{l-1}(u) = 0$$

It should be noted that for $l = 0$, the first root is given by:

$$J_{-1}(u) = J_1(u) = 0$$

As $w \longrightarrow \infty$, the characteristic equation gives:

$$J_l(u) = 0$$

The solutions for u in each mode therefore lie between the zeroes of $J_{l-1}(u)$ and $J_l(u)$. Every solution of the characteristic equation corresponds to a set of modes designated by LP_{lm} where, for $l \gg 1$, each set contains four modes.

If, however, n_1 and n_2 are not regarded as being equal and the boundary conditions are imposed, retaining terms linear in $(n_1 - n_2)$, it is found that the terms containing $(l + 1)\phi$ satisfy a different characteristic equation from those containing $(l - 1)\phi$, so that each LP_{lm} mode is degenerate and comprises two modes. One of these corresponds to the terms with $(l + 1)\phi$ and is designated $HE_{l+1,m}$, the other corresponds to the terms with $(l - 1)\phi$ and is designated $EH_{l-1,m}(l \neq 1)$ or TE_{om} and $TM_{om}(l = 1)$. This degeneracy is illustrated diagrammatically in Fig. 5 for the LP_{11} mode. This mode designation follows the scheme suggested by Bean, which depends on the relative contributions of E_z and H_z to the transverse components of the field at some arbitrary point. If the contribution from E_z is the largest, the wave is considered to be E-like and is designated EH, and so on.

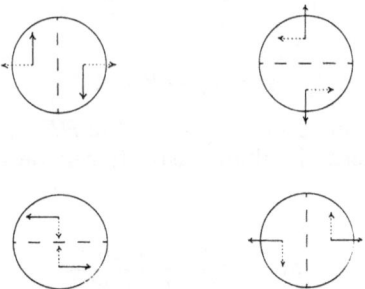

Fig. 5 Representation of the four possible field distributions in the L P_{11} mode. The electric vector is shown as a full line, the magnetic vector as a dotted line.

It can be shown that the characteristic equation (20) can be written in the form:

$$\frac{du}{dv} = \frac{u}{v}(1 - \kappa_l(w)) \tag{21}$$

where:

$$\kappa_l = \frac{K_l^2(w)}{K_{l-1}(w)\,K_{l+1}(w)} \tag{22}$$

An approximate solution to Eq. (21) can be obtained for all modes except $LP_{01}(= HE_{11})$ by replacing (22) by:

$$\kappa_l \approx 1 - (w^2 - l^2 + 1)^{-\frac{1}{2}}$$

and writing Eq. (13) as:

$$w \approx (v^2 - u_c^2)^{\frac{1}{2}}$$

where u_c is the cut-off value, given by $J_{l-1}(u_c) = 0$. This latter approximation is justified by the fact that u lies between the successive roots of Bessel functions for all values of w, and can be regarded as approximately constant. By substitution for these parameters, Eq. (21) can be written as:

$$\frac{du}{u} = \frac{dv}{v^2\left(1 - \left(\frac{s}{v}\right)^2\right)^{\frac{1}{2}}},$$

which can be solved to give:

$$u(v) = u_c \exp\left(\frac{\sin^{-1}\left(\frac{s}{u_c}\right) - \sin^{-1}\left(\frac{s}{v}\right)}{s}\right) \tag{23}$$

where:

$$s^2 = u_c^2 - l^2 - 1$$

These approximations do not apply to the HE_{11} mode, in which u, v and w approach zero simultaneously. It can be shown that for this mode:

$$u(v) = \frac{(1 + \sqrt{2})v}{1 + (4 + v^4)^{\frac{1}{4}}} \tag{24}$$

For large w (and v), i.e. far from cut-off, the value of u tends to the solution of $J_l(u) = 0$. If this is called u_∞, it can be shown that Eq. (23)

and Eq. (24) both reduce to:

$$u_v = u_\infty\left(1 - \frac{1}{v}\right)$$

far from cut-off.

A. Propagation Constant and Power Flow

The propagation constant, β, is dependent on the particular fibre configuration and it is more useful to use a normalized propagation constant, b, which is defined as:

$$b(v) = 1 - \frac{u^L}{v^2} = \frac{\frac{\beta^2}{k^2} - n_2^2}{n_1^2 - n_2^2} \tag{25}$$

If $n_1 - n_2 \ll 1$, this becomes:

$$b(v) = \frac{\beta/k - n_2}{n_1 - n_2} \tag{26}$$

The variation of $b(v)$ with v is shown in Fig. 6 for a number of L P modes.

During the transmission of an intensity-modulated light beam through a long optical fibre, the shape of the modulation envelope will change owing to the different propagation constants of the waveguide modes, which constitute the beam within the fibre. This

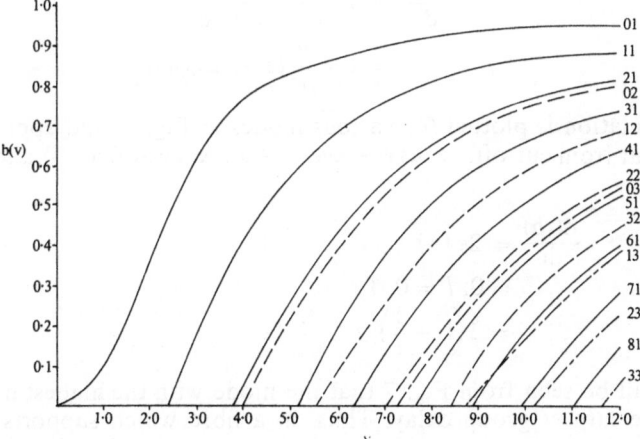

Fig. 6 Variation of the normalised propogation parameter, $b(v)$, plotted as a function of the normalized frequency, v, for various L P_{lm} modes. (After Gloge).

controlled by the group delay, τ, which is defined as:

$$\tau = \frac{L}{c} \frac{d\beta}{dk}$$

where L is the length of the fibre and c is the velocity of light *in vacuo*. Now, from Eq. (26):

$$\beta = k \left(b(n_1 - n_2) + n_2 \right)$$

If the dispersion of the core and sheath are equal then $(n_1 - n_2)$ is independent of k so that:

$$\frac{d\beta}{dk} = (n_1 - n_2) \frac{d(bk)}{dk} + \frac{d(n_2 k)}{dk}$$

$$= (n_1 - n_2) \frac{d(bv)}{dv} + \frac{d(n_2 k)}{dk} \qquad \text{from Eq. (13)}$$

Thus:

$$\tau = \frac{L}{c} \left\{ (n_1 - n_2) \frac{d(bv)}{dv} + \frac{d(n_2 k)}{dk} \right\} \qquad (27)$$

The second term in Eq. (27) describes the dispersion characteristics of the material and is the same for all modes. The first term accounts for the group delay within each waveguide mode. From Eqs. (25) and (21) it can be shown that:

$$\frac{d(bv)}{dv} = 1 + \frac{u^2}{v^2} - \frac{2u}{v} \frac{du}{dv}$$

$$= 1 - \frac{u^2}{v^2}(1 - 2\kappa_l(w)) \qquad (28)$$

This relation is plotted for various modes in Fig. 7, and approaches unity far from cut-off, i.e. as $v \longrightarrow w \longrightarrow \infty$. At cut-off $w = 0$, and $u = v$, so that:

$$\frac{d(vb)}{dv} = 2\kappa_l(w)$$

$$= 0; \, l = 0, 1$$

$$= 2\left(1 - \frac{1}{l}\right); \, l \geqslant 2$$

It will be seen from Fig. 7 that the mode with the highest order, l, has the largest group delay. Thus, in a fibre which supports many modes, the largest group delay will be shown by the highest-order mode which the fibre can support, and this mode will be close to cut-off. The lowest mode supported by the fibre will be far from cut-off,

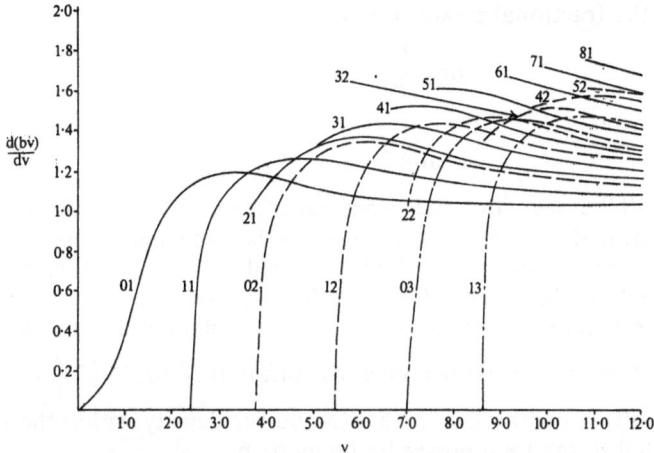

Fig. 7 Variation of normalized delay, $\frac{d(bv)}{dv}$, plotted as a function of the normalized frequency, v, for various L P_{lm} modes. (After Gloge)

so that the difference in group delays for these modes will be approximately:

$$\tau_l = \frac{L}{c}(n_1 - n_2)\left(1 - \frac{2}{l}\right)$$

Since, for large v, l_{max} approximates to v, the group spread is given by:

$$\tau_l = \frac{L}{c}(n_1 - n_2)\left(1 - \frac{2}{v}\right)$$

The power flow within the fibre can be obtained from the Poynting vector in the axial direction by integration over the fibre cross-section. This gives:

$$P_1 = \frac{Z_0}{n_1} \cdot \frac{1}{\kappa_l} \cdot \frac{\pi a^2}{2}\left(1 + \frac{w^2}{u^2}\right)E_l^2$$

for the power flow in the core, and:

$$P_2 = \frac{Z_0}{n_2}\left(1/\kappa_l - 1\right)\frac{\pi a^2}{2}E_l^2$$

for the power flow in the sheath. Since $n_1 \sim n_2$, the total power flow can be written as:

$$P = P_1 + P_2 = \frac{Z_0}{n_1}\frac{1}{\kappa_l}\frac{\pi a^2}{2}\frac{v^2}{u^2}E_l^2 \qquad (29)$$

Thus, the fractional power flow will be given by:

$$\text{core:} \frac{P_1}{P} = 1 - \frac{u^2}{v^2}\left(1 - \kappa_l\right) \Bigg\}$$

$$\text{sheath:} \frac{P_2}{P} = \frac{u^2}{v^2}\left(1 - \kappa_l\right) \Bigg\} \tag{30}$$

It will be seen that far from cut-off, i.e. $w \longrightarrow \infty$ the power is concentrated within the core, as would be expected. These fractional powers are plotted in Fig. 8 where it will be noted that, as cut-off is approached, the power of the two lowest modes ($l = 0, 1$) withdraws into the sheath whereas the higher modes still have a significant fraction of their power in the core at cut-off; in fact, for $l \geqslant 2, \frac{P_1}{P_2} = (l - 1)$ at cut-off. It can be shown that the power density within the system is related to the total power in the mode by:

$$p_1(r) = \kappa_l \frac{u^2}{v^2} \frac{2P}{\pi a^2} \left(\frac{J_l^2\left(\frac{ur}{a}\right)}{J_l^2(u)}\right) \cos^2 l\phi \Bigg\}$$

$$p_2(r) = \kappa_l \frac{u^2}{v^2} \frac{2P}{\pi a^2} \left(\frac{K_l^2\left(\frac{wr}{a}\right)}{K_l^2(w)}\right) \cos^2 l\phi \Bigg\} \tag{31}$$

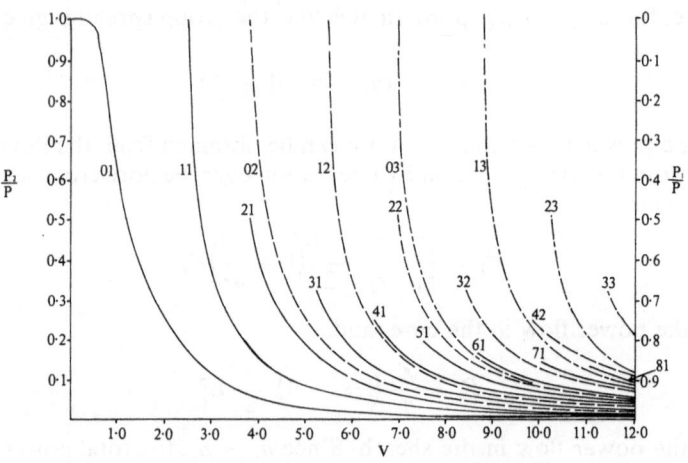

Fig. 8   Variation of the fractional power in the core $(\frac{P_1}{P})$ and that in the sheath $(\frac{P_2}{P})$, plotted as a function of the normalized frequency, v, for various L P_{lm} modes. (After Gloge)

where $p_1(r)$ and $p_2(r)$ are the power densities in the core and sheath respectively. By averaging Eq. (31) over ϕ the corresponding mean power densities are obtained:

$$\text{core: } \overline{p_1(r)} = \kappa_l \frac{u^2}{v^2} \frac{P}{\pi a^2} \left(\frac{J_l^2\left(\frac{ur}{a}\right)}{J_l^2(u)} \right)$$

$$p_2(r) = \kappa_l \frac{u^2}{v^2} \frac{P}{\pi a^2} \left(\frac{K_l^2\left(\frac{wr}{a}\right)}{K_l^2(w)} \right) \tag{32}$$

Of particular interest is the mean power density at the core – sheath interface since this is where power can be lost owing to defects at the interface. This is given by:

$$\overline{p(a)} = \kappa_l \frac{u^2}{v^2} \frac{P}{\pi a^2} \tag{33}$$

In Fig. 9 the variation of the normalized power density, $\mathscr{P}$ is plotted where:

$$\mathscr{P} = \frac{\pi a^2 \overline{p(a)}}{P} = \kappa_l \frac{u^2}{v^2}$$

For low-order modes ($l = 0, 1$), $\mathscr{P}$ tends to zero at cut-off, whilst for

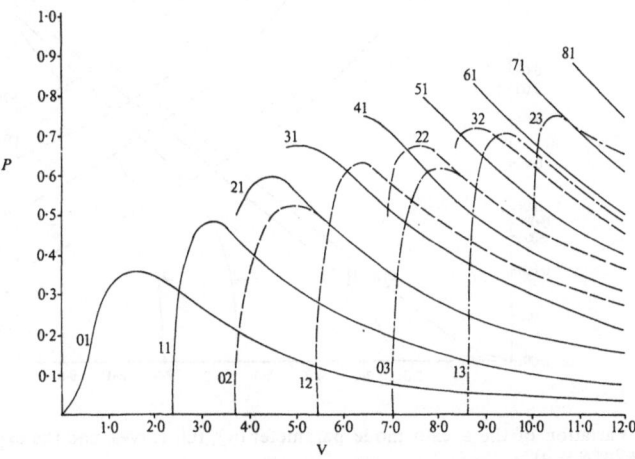

Fig. 9 Variation of the normalized interfacial power density ($\mathscr{P}$) plotted as a function of the normalized frequency, v, for various L P_{lm} modes. (After Gloge)

$l \geqslant 2$, $\wp$ tends to $\left(1 - \dfrac{1}{l}\right)$. Far from cut-off, $\wp$ tends to zero for all modes.

In a practical optical waveguide, the sheath thickness will be finite, but an analytical solution of this situation is not possible. However, some idea of the effects of a finite sheath thickness may be obtained by considering the power flow in the sheath for radii greater than the fibre radius R. In this case, R will be much greater than a and if the operating conditions are far from cut-off, the modified Hankel functions in Eq. (32) can be replaced by:

$$K_l(w) = \left(\frac{\pi}{2w}\right)^{\frac{1}{2}} \exp(-w)$$

The mean power density is then given by:

$$\overline{p_2(r)} = \kappa_l \frac{u^2}{v^2} \frac{P}{\pi a r} \exp\left(-2w\frac{(r-a)}{a}\right)$$

From this, it is a simple matter to prove that the fraction of the total power carried in the sheath, at radial distances greater than R, is given by:

$$f_R = \kappa_l \frac{u^2}{v^2 w} \exp\left(-\frac{2w(R-a)}{a}\right) \tag{34}$$

This fraction varies exponentially with distance from the interface and is shown in Fig. 10 for various modes. The power distribution in the

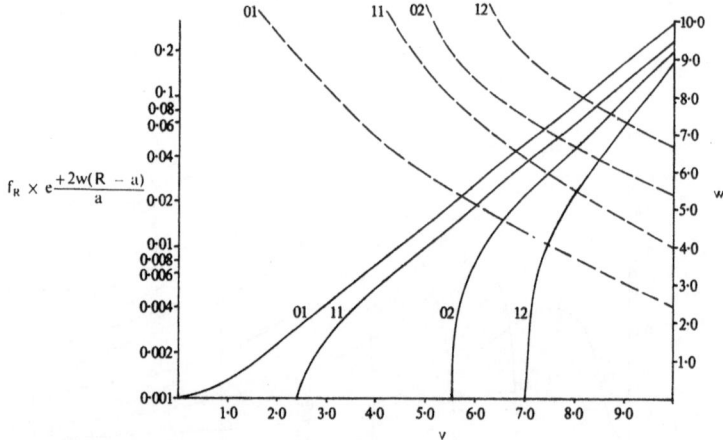

Fig. 10 Variation of the sheath mode parameter (w), full curves, and the expression $f_R \exp \dfrac{+2w(R-a)}{a}$, dotted curves, plotted as a function of the normalized frequency, v, for various L P_{lm} modes.

sheath close to cut-off can be obtained by replacing the modified Hankel functions by their approximations for small argument ($w \ll 1$) and setting $u = v$; the following relation is obtained:

$$\overline{p(r)} = \kappa_l \frac{P}{\pi a^2}\left(\frac{a}{r}\right)^l \tag{35}$$

This function decreases with distance from interface for all modes except the lowest mode whose field distribution is independent of radius, i.e. a plane wave.

B. Application to Large Fibres

With the larger fibres normally used in fibre optics, the preceding theory is too detailed to be usefully applied, owing to the large number of modes which are capable of being supported (Fig. 11). The following approach is a simpler method of dealing with such systems.

It has been shown that the core of light accepted by a fibre has a semi-angle θ, where:

$$\theta \approx \sin \theta = (n_1^2 - n_2^2)^{\frac{1}{2}}$$

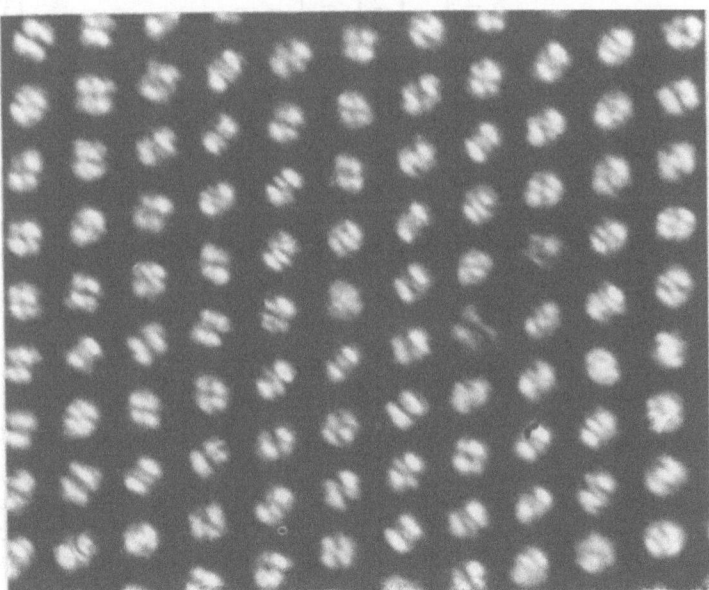

Fig. 11 A photomicrograph illustrating the complicated mode patterns set up in a bundle of conventional fibres (fibre diameter is 8μ). (Courtesy of Barr and Stroud Ltd)

The beam of light entering the fibre comprises free-space modes containing pairs of modes which are polarized perpendicular to each other. Each pair occupies a cone of solid angle $\left(\pi\dfrac{\lambda^2}{a^2}\right)$.[6] The total number of free-space modes accepted by the fibre is therefore given by:

$$N \approx 2\left(\frac{\pi a\theta}{\lambda}\right)^2$$

$$= \frac{v^2}{2} \qquad (36)$$

Since any particular mode order contains four sets, apart from the lowest one, the number of modes up to the nth order is given approximately by $4n$, if n is large. Thus, $v \sim (2n)^{\frac{1}{2}}$. At cut-off, $v = u_c$, so that cut-off for the $n$the mode occurs when:

$$u_c = (2n)^{\frac{1}{2}}$$

Since the value of u is restricted between the adjacent roots of successive Bessel functions, it can be replaced by u_c in the expression for propagation constant, giving:

$$b = 1 - \left(\frac{u_c}{v}\right)^2 = 1 - \frac{n}{N} \qquad (37)$$

and the group delay parameter can also be written as:

$$\frac{d(vb)}{dv} = 1 + \left(\frac{u_c}{v}\right)^2 = 1 + \frac{n}{N} \qquad (38)$$

where it has been assumed that most of the modes are far from cut-off i.e. $\kappa_l = 1.0$.

Similarly, the power density at the interface can be shown to be:

$$\overline{p(a)} = \frac{P}{\pi a^2}\left(\frac{u_c}{v}\right)^2 = \frac{P_n}{\pi a^2 N} \qquad (39)$$

The power flow in the sheath can also be shown to be:

$$P_2 = \frac{P_n}{N(2N - 2n)^{\frac{1}{2}}} \qquad (40)$$

If the source of light is incoherent, then every mode will be excited equally, so that average power distribution in the fibre is given by averaging the above relationships over all N modes. This yields for the total power density at the interface:

$$\left(\frac{\overline{p(a)}}{P}\right)_{\text{total}} = \frac{1}{\pi a^2}\int_0^N \frac{n\,dn}{N} = \frac{1}{2\pi a^2} \qquad (41)$$

which depends solely on the fibre radius. The average power in the sheath is given by:

$$\left(\frac{P_2}{P}\right)_{total} = \frac{4}{3}N^{-\frac{1}{2}} = \frac{4\sqrt{2}}{3v} \tag{42}$$

For example, an optical fibre with a core radius of 25 μm and an N.A. of 0.3 will support over 3000 modes at a wavelength of 0.6 μm which justifies the use of the geometric approach in the earlier chapters. Approximately 2% of the power is propagated in the sheath and the group delay over one kilometre is 100 ns.

III. PRACTICAL OPTICAL-FIBRE WAVEGUIDES

The main area of use for a practical optical waveguide will be in the communications field.[7] The present communications network is formed from coaxial cable or microwave radio relay systems and demand is doubling every six years or so. The increasing demand for data transmission and the projected introduction of a videophone system may further accelerate the growth of this network. It is felt that development of existing systems will not keep pace with the growth and the optical waveguide offers an alternative system with a much larger bandwidth. This system would be economically attractive if the repeater stations could be spaced at distances greater than 1 km, which requires an optical fibre with an attenuation of less than 20 dB/km. The power source for the carrier would ideally be solid-state, the GaAs laser being the most obvious choice at present; this operates at about 0.9 μm.

The attenuation of existing optical glasses is an order of magnitude higher than the required figure for the optical waveguide and the attenuation of normal optical fibres is two or three times as large as that of the bulk glass, figures of 600 dB/km being typical. From this it is obvious that new glasses and manufacturing techniques will have to be developed before such low attenuations can be achieved. In fact, work has already started on this in several countries and glasses have been developed with bulk attenuations of less than 20 dB/km. The requirements for such a glass will be outlined in the next section.

Since the optical carrier will be modulated at very high frequencies, it is essential that the waveguide shall operate in a single mode only (Fig. 12.). The simplest way of achieving this is to design the fibre so that only one mode can be supported and a possible approach to this is to choose the LP_{01} mode (HE_{11}), which is the lowest-order mode. For the next higher mode (LP_{11}), $u_c = 2.405$ and if the fibre is designed so

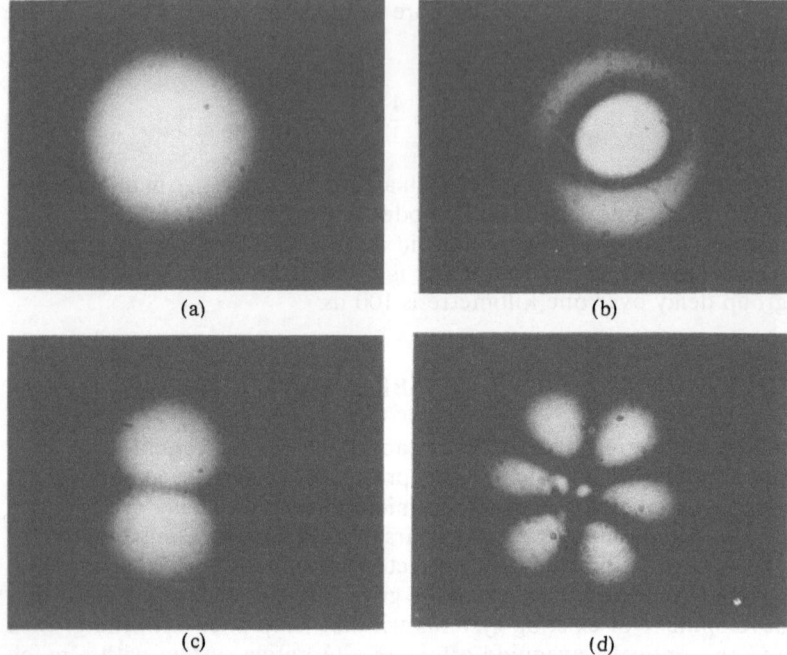

Fig. 12 Photographs of the radiation patterns of optical waveguide modes.
(a) HE_{11} mode in a 1.4 μ fibre.
(b) HE_{12} mode in a 5.3 μ fibre.
(c) TE_{01} + HE_{21} modes in a 3.0 μ fibre.
(d) A mixture of low order modes in a 5.3 μ fibre, possibly EH_{21} + HE_{41}.

The fine interference fringes are caused by dust on the lens used to take these photographs and by the lens itself. (Courtesy of S.T.L. Ltd Harlow)

that $v < 2.405$ then only the HE_{11} mode will be supported. In fact, it will be seen from Fig. 6 that this mode has no cut-off value and will be propagated for very small v. A small value of v can be achieved by reducing the fibre diameter and this mode has been seen in fibres as small as 0.1 μm corresponding to a v-value of about 0.18. However, these small diameters create manufacturing problems and also give a low propagation constant.

A more realistic design, which is being studied, consists of a 2 μm core of glass with a refractive index of 1.52, in a sheath of refractive index 1.50, which gives a v-value of about 1.6 at 0.9 μm. The propagation constant is $1.504 k$ which indicates that most of the light is carried in the sheath, in fact about 55% from Fig. 8. In order to minimize the effects of the finite sheath thickness, the total fibre diameter must be as large as possible but the fibre must be fairly flexible to permit handling, and for this reason the total diameter must be kept below 100 μm, 50 μm being a better value for ease of handling. From Fig. 10 is can be shown that the fraction of the total power which should theoretically lie outside these diameters is respectively 1.6×10^{-36} and 4×10^{-18}. This means that the effect of the finite sheath thickness will be minimal, even at the lower diameter.

A. The Manufacture of Optical Waveguides

As was indicated in the previous section, existing glasses and manufacturing techniques are not suitable for this type of optical fibre and the development of new materials and processes is under way at present.[8]

The loss of light in glass can be divided into two components, namely scattering and absorption losses. The scattering losses in optical glasses tend to be low, varying from 0.6 db/km to 7.0 db/km, where the higher losses are found in glasses containing the heavier elements, e.g., lead. It seems then that the scattering losses in the bulk glass are acceptable, particularly for glasses containing light elements, and that the major loss is due to absorption. The main contribution to the absorption in glasses comes from the transition metals (Cu^{2+}, Fe^{2+}, Ni^{2+}) and the concentrations of the cupric and ferric ions must be kept below 0.1 ppm and 0.01 ppm respectively to achieve a glass with the desired transmission. Another contaminant is platinum which is dissolved from the crucible during the melting process and can cause both scattering and absorption losses. If the platinum particles are sufficiently small, only absorption will take place, and under these conditions 1 ppm of platinum will give rise to an attentuation of 10^3 dB/km. It is obvious from this that the glasses should not be melted in platinum and a suitable alternative would be pure alumina; the results of melting the same glass in platinum and alumina are shown in Fig. 13. There is also some evidence that

	Crucible	
	Platinum	Alumina
Analysis (p.p.m.) Pt	6.7	0.5
Fe	0.7	5.9
Cu	2.2	1.9
Measured loss (dB/km)	235	100

Fig. 13 Typical trace element analyses and loss measurements for glass melted in platinum and alumina crucibles. The base glass composition was 70% Si O_2, 24% $Na_2 O$, 6% CaO (by weight). (Courtesy of Dep. of Glass Technology, Sheffield University)

absorption is due to a harmonic of the hydroxyl absorption at 2.74 μm from water dissolved in the glass, and this loss has been calculated to be 1.25 dB/km for 1 ppm of water. The glasses quoted in Fig. 9 were prepared, using conventional techniques, from raw materials with less than 0.1 ppm of both copper and iron, so additional contamination is being introduced during the processing. In spite of these difficulties, glasses have been made with total attentuations less than 20 dB/km.

The increased attenuation which occurs when the bulk glasses are drawn into fibres seems to be attributable mainly to an increase in the scattering loss, since it is unlikely that the processing would alter the absorption of the glasses to this extent. The glasses which are used in these fibres are very stable at elevated temperatures, so it would appear that the increased scatter occurs mainly at the interface, and is due to either contamination or mixing of the two glasses which would cause slight devitrification or phase-separation at the interface. Both these effects would significantly increase the scattering losses and it is therefore advisable to use a design of fibre with a low power density at the core – sheath interface. This is the case for the HE_{11} mode with low v-values.

Methods of decreasing the contamination at this interface have been investigated and two of these are showing promise. In the first, the fibre is drawn using a double-crucible technique, which was developed for the manufacture of optical fibres for use in the infra-red region of the spectrum and is discussed in detail in the next chapter. This technique enables fibres to be drawn directly from the melt, which means that the interface is formed during the drawing process and is not produced by the fusion of existing surfaces, as in the conventional fibre-drawing technique. For this reason, the interface should be free from contamination owing to contact with the atmosphere. However, the operating temperature in the double-crucible technique is much higher than in the conventional process and this will encourage the mixing of the glasses at the interface, which will increase the effect of any phase separation.

The second approach to this problem is to produce a fibre without a definite interface and is achieved by drawing a fibre with a refractive index which decreases uniformly with distance from the axis. This variation is produced by immersing a glass rod in a bath of molten salt and allowing the heavy ions in the rod to be replaced by the lighter ions of the salt by a diffusion process. Obviously, this exchange will take place more rapidly in the outer layers of the rod producing a gradual increase in refractive index towards the centre. The theory of operation of this type of fibre will be outlined in the next chapter.

REFERENCES

1. S. A. Schelkunoff, *Electromagnetic Waves*, D. Van Nostrand Company, Inc., Princeton (1943), Ch. 10
2. D. Hondros and P. Debye, *Ann. Physik* **32**, 465 (1910)
3. E. Snitzer, *J. Opt. Soc. Am.* **51**, 491 (1961)
4. N. S. Kapany, J. J. Burke and C. C. Shaw, *J. Opt. Soc. Am.* **53**, 929 (1963)
5. D. Gloge, *Appl. Opt.* **10**, 2252 (1971)
6. G. Toraldo di Francia, *J. Opt. Soc. Am.* **59**, 799 (1969)
7. K. C. Kao and G. A. Hockham, *Proc. Instn Elec. Engrs* **113**, 1151 (1966)
8. B. Scott and H. Rawson, Proc. Symposium on Electrotechnical Glasses, Society of Glass Technology, Sheffield, (Sept. 1970)

CHAPTER 10

Miscellaneous Topics

As the technology of fibre optics developed and the manufacturing techniques became established, there arose an interest in the application of these techniques in areas outside the generally accepted field of fibre optics, viz the guidance of visible light by channels large compared to the wavelength of light being used. In this chapter, a few of the more interesting ones will be discussed; at the moment these are of slight commercial importance owing either to a high manufacturing cost or to the lack of a suitable market, but there is no doubt that this situation will alter in the future as the development of the technology progresses.

The areas to be covered are the use of fibre optics outside the visible spectrum, the properties of the optical fibre as an active system, and the properties of an optical fibre with a continuous radial change in refractive index instead of the normal discrete step at the core – sheath interface.

I. FIBRE OPTICS OUTSIDE THE VISIBLE SPECTRUM

The spectral transmission of an optical fibre is determined by the raw material of the core and sheath. The materials which are normally used are optical-quality glasses that cover a spectral range larger than the visible portion of the spectrum; the exact extent of the range is determined by the overall length of the bundle but it covers approximately 0.35 μm to 1.5 μm. Within this range, therefore, the normal optical fibre is perfectly adequate and bundles of such fibres can be used with silicon photo-detectors and Ga – As photo-emitters, which peak in the near infra-red (around 0.9 μm) and with phosphors which peak in the near ultra-violet. If it is desired to work outside this range then different raw materials must be used.

A. Ultra-violet Fibre Optics

The obvious material to use for the manufacture of optical fibres to be employed in the part of the spectrum below 0.35 μm is pure silica. Unfortunately, this has a very low refractive index, $nD = 1.458$, and it is impossible to find a suitable sheath material in the range of suitable glasses. However, a number of plastics, particularly silicone resins, have a suitably low refractive index and can be used to coat the drawn silica fibre. The normal method of applying the sheath is to pass the fibre through a bath containing the resin. Any absorption in the plastic is of secondary importance since the overall path-length of a typical ray in the sheath is very small in the lengths of bundle normally used.

Fibres of this type can be assembled into bundles in very much the same way as normal optical fibres, although greater care is required during handling to avoid damaging the plastic sheath. At present, these fibres are only available in the form of light guides and find application in fluorescence and spectro-photometric work in the ultra-violet, The variation of transmission with wavelength of a typical u.v. light guide is shown in Fig. 1.

The intrinsically low absorption of pure silica is also of interest in the visible and near infra-red portions of the spectrum, where very long lengths of optical fibre are desired. However, the plastic sheath used in the above design of fibre would reduce the overall transmission considerably in the lengths that are of interest. A practical solution is to use the silica as a sheath material and a modified silica as the core, where this modification increases the index of the resultant glass without significantly increasing the absorption of the resulting fibre.

B. Infra-red Fibre Optics

Since normal optical glass covers the near infra-red portion of the spectrum adequately, the main interest in this area has been centred on the parts of the spectrum known as 'atmospheric windows.' These are the portions of the spectrum in which little or no solar energy reaches the earth's surface owing to atmospheric absorption, and can be

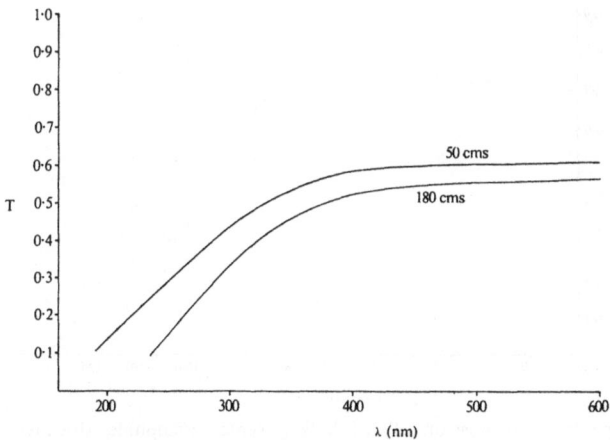

Fig. 1 Spectral transmission (T) of UV light guides of lengths 50 cm and 180 cm. (Courtesy of Schott and Gen.)

broadly defined as covering the regions from 3.5 μm to 5.0 μm and 8.0 μm to 14.0 μm. This interest derives from the use of these 'windows' in the radiometry of guided missiles since, in these regions of the spectrum, the exhaust from a jet or rocket engine would appear against a 'black' background.

There are a few glasses which have the desired transmission characteristics and some of these have been successfully drawn as fibres.[1] The most suitable material is arsenic trisulphide, which is a naturally vitreous material with a large refractive index, around 2.5 in the regions of interest. The spectral transmission of a bulk sample of arsenic trisulphide is given in Fig. 2. The material has a very low softening point ($T_g = 250°$C) and a relatively high expansion coefficient ($\alpha = 25 \times 10^{-6}/°$C) so that the sheath glass has to be a modified arsenic trisulphide in order to obtain similar values for this glass. A suitable modifier is sulphur, which lowers the refractive index sufficiently without drastically altering any of the other important properties.

Fibres can be manufactured from these materials using the 'rod and tube' method if tubing can be made from the modified glass, although conventional tube drawing techniques cannot be used owing to the high thermal expansion of the glass, which causes the drawn tube to shatter on cooling. However, a simple manufacturing technique has been developed which does not require a drawing process and has the further advantage of requiring only small quantities of raw materials. This technique can be used with any glass which possesses the following three properties:

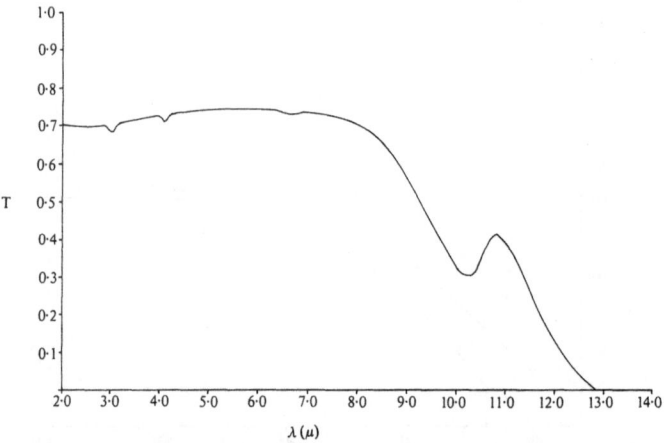

Fig. 2   Spectral transmission (T) of bulk Arsenic Trisulphide, thickness 6.3 mm. (Courtesy of Barr and Stroud Ltd)

1. A low softening point.
2. A high thermal expansion coefficient.
3. Inability to 'wet' oxide glasses.

Since there are a number of glasses which have these properties and also transmit well in the infra-red, this technique is of general interest.

A measured quantity of the glass is placed inside a tube of a low-expansion oxide glass, e.g., fused silica, or even a borosilicate glass if the melting point of the base glass is low enough. The tube is evacuated, sealed at both ends and mounted horizontally in a furnace, as shown in Fig. 3, which is then taken up in temperature till the base glass is fluid. The furnace is held at this temperature and the tube is slowly rotated at such a speed that the molten glass coats the inside of the tube uniformly. When this condition is reached the furnace temperature is lowered to the annealing point of the glass, while still rotating the tube, and then brought to room temperature at a rate chosen to avoid shattering the glass. When the base glass sets, the high differential expansion will cause this tube to shrink away from the oxide-glass tube, owing to the 'non-wetting' property of the glasses. The oxide-glass tube is removed from the furnace and opened so that the base-glass tube can be withdrawn.

With arsenic trisulphide, the required furnace temperature is around 450°C so that a Pyrex tube can be used to contain the base glass. The resulting tube has a 'flame-polished' bore which is an ideal surface for the subsequent fibre drawing operation. Precautions must be taken during the drawing of fibres from this material since poisonous fumes can be produced. A typical furnace for the drawing of these

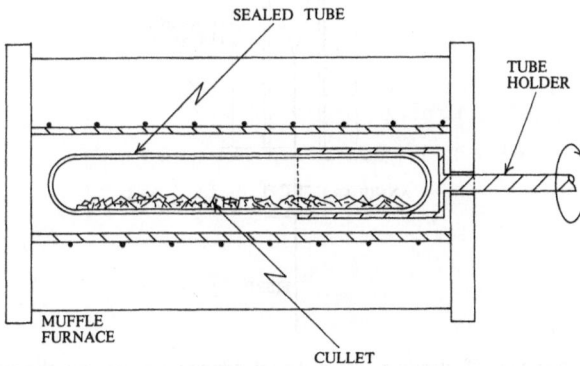

Fig. 3   Furnace set up used in the manufacture of tubing from a modified Arsenic Trisulphide glass.

fibres is shown in Fig. 4; an inert gas is bled into the base of the furnace and an extraction system removes the gases from the top of the furnace.

An alternative manufacturing technique can be used for glasses which cannot be treated in the above manner. This is known as the double-crucible technique and the basic principles are outlined below.

The double crucible consists of a central pot, with an orifice in the base, which holds the core glass, and a concentric outer pot, which holds the sheath glass and also has an orifice in the base; again the materials used will be silica or a borosilicate glass. The whole assembly is mounted vertically in a furnace with the orifices in line, as shown in Fig. 5 and, when the correct temperature conditions are attained, a fibre can be drawn from the lower orifice, which will consist of a core and sheath if the upper orifice is sufficiently close to the lower one. The thickness of the sheath depends on a number of factors including the furnace temperature, pressure head of glass, the orifice diameters and their separation, whilst the diameter of the drawn fibre depends on the furnace temperature and the drawing rate. For safety reasons the assembly is contained within an enclosure with an inert atmosphere.

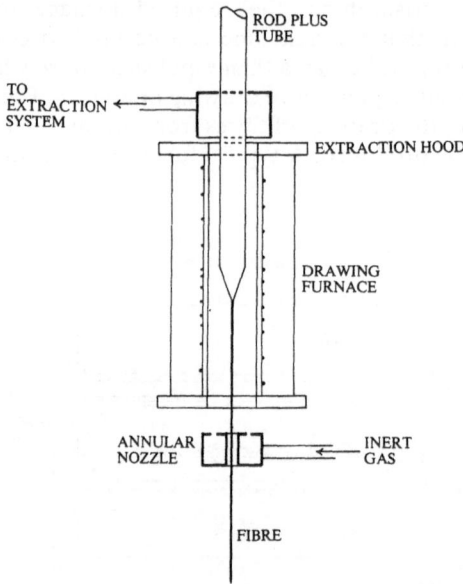

Fig. 4 Arrangement of bleeding system to introduce inert gas into the muffle of the drawing furnace. This is used when fibres are drawn from glasses which emit toxic fumes on heating.

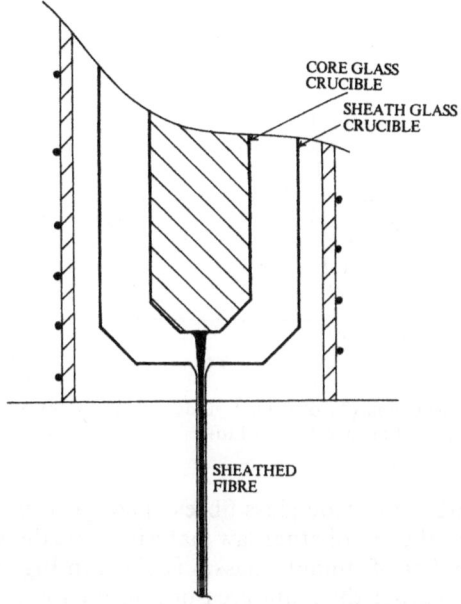

CORE GLASS
CRUCIBLE

SHEATH GLASS
CRUCIBLE

SHEATHED
FIBRE

Fig. 5   Illustration of the double crucible technique for the manufacture of sheathed fibres.

Because of the large number of parameters influencing the sheath thickness this process is a much more difficult one to control than the normal 'rod and tube' method; however it does offer two significant advantages over this technique. First, the glass can be used in cullet form, which reduces the amount of raw material required, and secondly, the core – sheath interface is formed during the drawing process and not by the fusion of two previously formed surfaces, as in the 'rod and tube' technique. This means that the interface should be very much cleaner in the double-crucible technique and there should be no chemical contamination of the glass on either side of the interface. The resulting fibre should therefore be of a much higher quality than that produced by the normal technique and this may become significant in the future when applications involving very long fibres, and therefore a large number of reflections, become commercial propositions. This argument applies to the entire spectral range covered by fibre optics, although with optical glasses the crucibles will have to be made from either refractory materials or platinum.

The spectral transmission of a bundle of fibres with an arsenic trisulphide core and a modified arsenic trisulphide sheath is shown in Fig. 6. Such fibres can be assembled into the complete range of com-

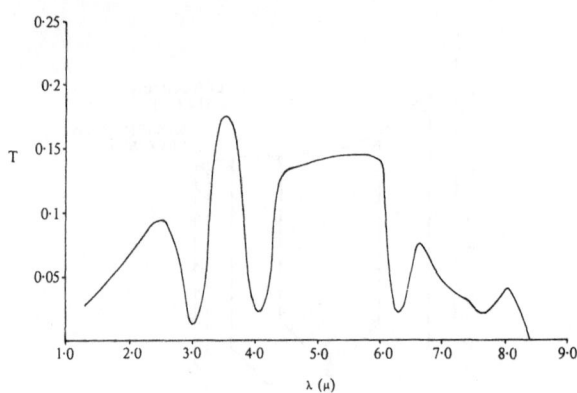

Fig. 6   Spectral transmission (T) of a light guide (30 cms long) made from Arsenic Trisulphide. (Courtesy of Barr and Stroud Ltd)

ponents available with oxide-glass fibres. The spectral range covered can be altered by the use of other raw materials, and the spectral transmission of a number of suitable glasses is given in Fig. 7.

As was stated earlier the main area of interest for such fibres is in the radiometry of guided missiles where their use permits a flexibility in the radiometer design not achievable by conventional means.

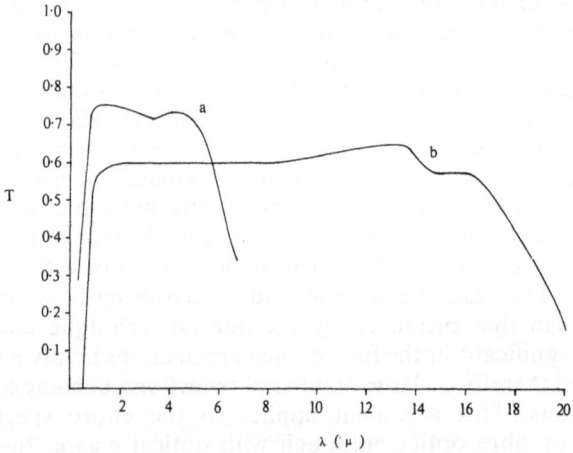

Fig. 7   Spectral transmission (T) of glasses suitable for infra-red optical fibres.
(a) Tellurick glass
(b) Arsenic triselenide
(Courtesy of Barr and Stroud Ltd)

II. ACTIVE OPTICAL FIBRES

The normal optical fibre is a passive device and functions solely as a guiding means for electro-magnetic energy. However, this need not be the case and two types of optical fibre have been developed in which an active role is added to the optical performance of the fibre. In the first of these, the core of fibre is made from a glass which exhibits 'lasing' action and the resulting system acts as its own resonant cavity, resulting in an optical-fibre laser.

Optical-Fibre Laser

Although many ions can be made to exhibit 'lasing' action, problems are encountered in trying to use glass as the host material since the lack of periodicity in the vitreous network seriously pertrubs the energy levels in the active ion. In 1961, Snitzer[2] demonstrated 'lasing' action at room temperature in a glass using the trivalent neodymium ion emitting at 1.06 μm. The closed shells in the outer electronic orbits of the rare earths effectively shield the ion from the perturbing influence of the host network. A barium crown glass is the most suitable choice of a host material since this exhibits the longest lifetime for fluorescent decay, and an unsheathed fibre of this material will act as a laser with the end-faces of the fibre defining the resonant cavity. With this configuration, the reflection at the end-faces is provided by the Fresnel reflectivity, which is low, and a relatively long length of fibre must be used to give reasonable threshold values. For this reason, the fibre laser is normally sheathed with a normal glass of lower refractive index to provide a protected reflecting interface, as in normal optical fibres. This sheath can also lower the 'lasing' threshold, since it acts as a cylindrical lens which focusses more of the pumping light into the core. Further, by making the difference in refractive index small, the mode selection within the system can be improved.

The set-up for the operation of an optical-fibre laser is illustrated in Fig. 8 where it will be seen that the fibre is of sufficiently small diameter to be wound helically around the flash-tube. This obviously improves the coupling between the pump source and the laser and results in very low operating thresholds ($\approx$ 1 joule). Thus, continuous lasing action can be obtained by wrapping the fibre around a filament bulb with a power of 1–2 kW. As the pumping level is increased, the bandwidth of the emitted light increases and the output light levels exhibit a periodic fluctuation owing to cyclic effects in the stimulated emission processes.

Since the actual volume of active glass in the system is small, the optical fibre laser does not have a high power output and does not offer competition to the normal rod laser. In addition, the mode

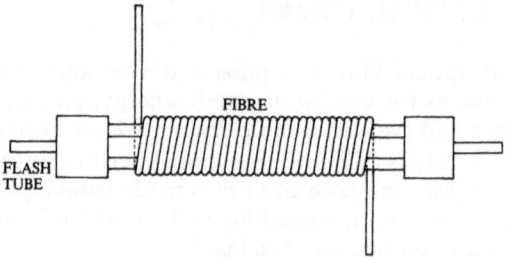

Fig. 8   Pumping arrangement for an optical fibre laser.

selection which can be achieved is significantly less than in the 'rod' system. The main interest in this device is in its potential use as a light amplifier, in which role the optical fibre is operated below the 'lasing' threshold and a light beam, of the appropriate wavelength, is amplified by passage through the fibre owing to the stimulated emission of photons by the beam. To prevent 'lasing', the single-pass gain must be less than the reciprocal of the Fresnel reflectivity, which limits the gain to around × 20 in the fibre with normal end-faces. However, if the end-faces are angled so that the reflected light is not internally reflected at the core – sheath interface, then very low effective reflectivities can be achieved (Fig. 9). Using this technique, gains of up to × 10⁵ have been achieved.[3]

If the amplifier is operated sufficiently close to the 'lasing' threshold, then the additional pumping due to the passage of the light beam can initiate 'lasing' action. This illustrates the use of an optical fibre laser as an optical trigger where a beam of low intensity can be used to initiate a beam of much higher intensity.

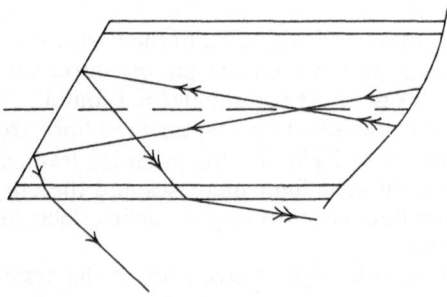

Fig. 9   Elimination of the effects of Fresnel reflection in an optical fibre amplifier.

Luminescent Fibres.

The optical performance of the optical fibre was outlined in Chapter 2, where it was assumed that the source of light was external to the system. If the core material, on the other hand, is made from a luminescent material, then the passage of an X-ray or high-energy particle will stimulate emission of light within the fibre.[4] It is obvious that this is an identical situation to the case of an optical fibre embedded in a material with the same refractive index as the core, so that the effective N.A. can be derived from Eq. (7) of Chapter 2 by substituting n_1 for n_0 giving:

$$\text{N.A.} = \frac{1}{n_1}(n_1^2 - n_2^2)^{\frac{1}{2}} \tag{1}$$

Since the light will be emitted in all directions the fraction of light accepted by the fibre will be given by $1 - \frac{n_2^2}{n_1^2}$, in each direction along the fibre axis.

The overall efficiency of transmission for this light can then be calculated as outlined in Chapter 2, if the absorption coefficient and core – sheath reflectivity is known. Such luminescent fibres can be made from suitable plastics, glasses or liquids, the fibres in the latter instance being made by containing the liquid in glass tubes with a refractive index less than that of the liquid. A novel detection system has been developed using fibres of this type which gives immediate stereoscopic display of the path of a high-energy particle through the system.[5] This is illustrated in Fig. 10 and consists of single coherent layers of luminescent fibres which are stacked alternately at right angles to each

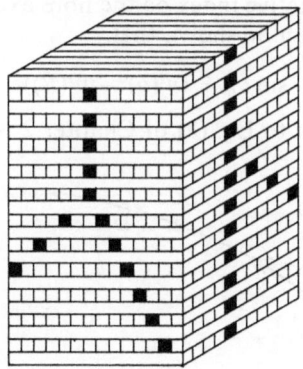

Fig. 10 An array of coherent layers, with luminescent fibres, used to indicate the tracks of a nuclear event in three dimensions. Activated fibres are shown in black.

other, so that the projection of the track on two orthogonal planes is displayed.

III. GRADED-INDEX FIBRES

A novel type of optical fibre has been developed recently in which guidance is achieved by decreasing the refractive index uniformly with distance from the axis. One reason for this development has already been discussed in Chapter 9, in connection with the scattering losses introduced by defects at the core – sheath interface. The path of a light ray through such a fibre will not, in general, be a straight line and, in the case of guidance, will take the form of a periodic oscillation about the fibre axis. This is illustrated in Fig. 11 and it can be shown that a generalized Snell's law can be applied to this situation, giving:

$$\cos \phi_1 \, n(r_1) = \cos \phi_2 \, n(r_2) = \cos \phi \, n(r) \qquad (2)$$

where ϕ is the angle which the ray makes with the axis, and $n(r)$ defines the radial variation in refractive index.

If guidance is achieved, then at some point in its path the ray will be parallel to the axis, and this will correspond to the maximum radial excursion of the ray, which will be designated by R. This can be regarded as the point of reflection and corresponds to the core–sheath interface in the conventional fibre. If a ray is incident at the end-face of this fibre with an internal angle ϕ_0, see Fig. 12, then from Eq. (2) we have:

$$\cos \phi_0 \, n(o) = \cos \phi \, n(r) = n(R)$$

where $n(o)$ is the refractive index on the fibre axis. If the external angle of incidence is θ_0 it can be shown that:

$$\sin \theta_0 = (n^2(o) - n^2(R))^{\frac{1}{2}} \qquad (3)$$

The similarity of this to Eq. (8) of Chapter 2 is obvious and Eq. (3)

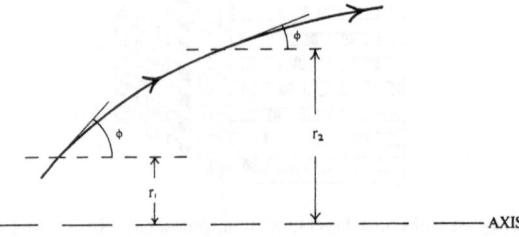

Fig. 11 Generalized path of a light ray in a medium where the refractive index, $n(r)$, is a function of distance (r) from an axis.

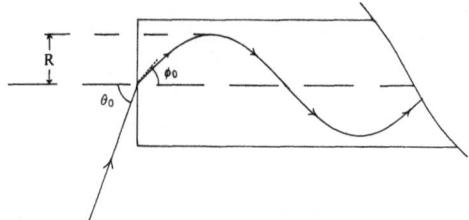

Fig. 12   The path of a ray of light in a graded index fibre.

gives the acceptance angle for a graded index fibre of radius R. Thus the difference in refractive index to achieve a given N.A. is the same for both fibres.

The condition for guidance is simply that the refractive index decreases monotonically with distance from the axis. However, a further condition must be fulfilled if the fibre is to support waveguide modes. This is that the mean axial velocity of any ray over one oscillatory period should be constant, which is equivalent to saying that the optical path lengths of the rays are constant. Therefore, a fibre which satisfies this condition will also be capable of forming an image of an object in close contact with its end-face, where the object/image distance equals the periodic length. This condition can be expressed in the following manner:

$$\int_s n(r)\,\mathrm{d}s = \text{constant} \tag{4}$$

where $\mathrm{d}s$ is an element of length along the ray, at a distance r from the axis and the integration is performed over a periodic length. To evaluate this integral, Eq. (2) is squared and the following substitution made: $\dfrac{1}{\cos^2\phi} = 1 + \tan^2\phi = 1 + \left(\dfrac{\mathrm{d}r}{\mathrm{d}z}\right)^2$ where z is the co-ordinate along the fibre axis. This gives:

$$\left(\frac{\mathrm{d}r}{\mathrm{d}z}\right)^2 = \frac{n^2(r)}{\cos^2\phi_0\, n^2(o)} - 1 \tag{5}$$

where ϕ_0 is the axial angle at $r = 0$ and $n(o)$ is the refractive index on the axis. Now:

$$\mathrm{d}s^2 = \mathrm{d}r^2\left(1 + \left(\frac{\mathrm{d}z}{\mathrm{d}r}\right)^2\right)$$

$$= \frac{\left(\mathrm{d}r^2\,\dfrac{n^2(r)}{n^2(o)\cos^2\phi_0}\right)}{\left(\dfrac{n^2(r)}{n^2(o)\cos^2\phi_0} - 1\right)}, \text{ from Eq. (5)}$$

Therefore

$$ds = \frac{n(r)\ dr}{n(o)\ \cos\phi_0 \left(\dfrac{n^2(r)}{n^2(o)\ \cos^2\phi_0} - 1\right)^{\frac{1}{2}}}$$

Then Eq. (4) becomes:

$$\int_{s'} \frac{n^2(r)\ dr}{n(o)\ \cos\phi_0 \left(\dfrac{n^2(r)}{n^2(o)\ \cos^2\phi_0} - 1\right)^{\frac{1}{2}}} = \text{constant}$$

This can be written as:

$$4\int_0^R \frac{n^2(r)\ dr}{n(o)\ \cos\phi_0 \left(\dfrac{n^2(r)}{n^2(o)\ \cos^2\phi_0} - 1\right)^{\frac{1}{2}}} = \text{constant} \qquad (6)$$

where R is the maximum value of r.

An obvious form for the path of the ray is a sinusoidal oscillation, i.e.

$$r = R \sin\alpha z, \text{ where } \alpha = \frac{2\pi}{L},$$

L being the periodic length. In this case: $\dfrac{dr}{dz} = R\,\alpha\,\cos\alpha z$. Now, at $z = 0$, $\tan\phi_0 = R\,\alpha$, and it can be shown that, from Eq. (5):

$$n^2(r) = n^2(o)(1 - \cos^2\phi_0\ \alpha^2\ r^2) \qquad (7)$$

Using these, Eq. (6) becomes:

$$\int_{s'} n(r)\ ds = 4\int_0^{\frac{\tan\phi_0}{\alpha}} \frac{n(o)(1 - \alpha^2 r^2 \cos^2\phi_0)}{\sin\phi_0(1 - \alpha^2 r^2 \cot^2\phi_0)^{\frac{1}{2}}} dr$$

$$= \frac{4n(o)}{\alpha\ \cos\phi_0} \int_0^{\frac{\pi}{2}} (1 - \sin^2\phi_0\ \sin^2\theta) d\theta$$

where $\sin\theta = \alpha\ r\ \cot\phi_0$

This can be immediately integrated to give the optical path length:

$$\int_{s'} n(r)ds = \frac{2\pi\ n(o)}{\alpha} \cdot \frac{(1 - \frac{1}{2}\sin^2\phi_0)}{\cos\phi_0} \qquad (8)$$

This is obviously not constant for all rays, but if the consideration is restricted to paraxial rays,[6] then $\cos\phi_0 \approx 1$ and $\sin\phi_0 \ll 1$, thus Eq.

(8) becomes:

$$\int_{s'} n(r)ds = \frac{2\pi\,n(o)}{\alpha} = n(o)L$$

Therefore, with this refractive index distribution, paraxial rays emitted from a point on the axis are brought to a focus at a distance L along the axis. For the paraxial case, the refractive index varies according to the following, from Eq. (7).

$$n(r) = n(o)(1 - \alpha^2 r^2)^{\frac{1}{2}} \approx n(o)(1 - \tfrac{1}{2}\alpha^2 r^2) \tag{9}$$

A distribution of refractive index can be found which focusses all rays from a point on the axis;[7] this distribution is:

$$n(r) = n(o)\,\text{sech}\,(\alpha r)$$

$$= n(o)\left(-1\tfrac{1}{2}\alpha^2 r^2 + \frac{5}{24}\alpha^4 r^4 - \frac{61}{720}\alpha^6 r^6 + \ldots\ldots\right.$$

$$\left.\ldots\ldots + \frac{E_n}{(2n)!}(\alpha r)^{2n} + \ldots\ldots\right) \tag{10}$$

where E_n are the Euler numbers. Clearly Eq. (10) reduces to Eq. (9) for $\alpha r \ll 1$. If this expression is substituted in Eq. (6) we have:

$$\int_{s'} n(r)ds = 4 \int_0^R \frac{n(o)\,\text{sech}^2\,(\alpha r)dr}{(\text{sech}^2\,(\alpha r) - \cos^2\,\phi_0)^{\frac{1}{2}}}$$

and, using the substitution $x = \tanh\,(\alpha r)$, this becomes:

$$\int_{s'} n(r)ds = \frac{4n(o)}{\alpha} \int_0^{\tanh(\alpha R)} \frac{dx}{(\sin^2\,\phi_0 - x^2)^{\frac{1}{2}}}$$

$$= \frac{4n(o)}{\alpha}\sin^{-1}\left\{\frac{\tanh\,(\alpha R)}{\sin\,\phi_0}\right\} \tag{11}$$

Now, from Eq. (2):

$$\cos\,\phi_0 = \frac{n(R)}{n(o)} = \text{Sech}\,(\alpha R)$$

So that,

$$\tanh\,(\alpha R) = (1 - \text{sech}^2\,(\alpha R))^{\frac{1}{2}} = \sin\,\phi_0$$

Therefore, Eq. (11) becomes:

$$\int_{s'} n(r)ds = \frac{2\pi\,n(o)}{\alpha} = n(o)L$$

for any value of ϕ_0.

The above argument will apply to rays from an off-axis point provided that these rays pass through the axis, i.e. meridional rays. Thus, for meridional rays, the conjugate planes for this optical system are any two planes normal to the axis which are separated by a distance L.

Unfortunately, this argument does not apply to skew rays[8] and, to illustrate this, the behaviour of a particular class of skew ray will be considered, namely those which follow a helical path through the fibre. Rays which are in this class maintain a constant distance from the axis so that the refractive index is constant. In order to satisfy the constancy of path length, the ray must travel a distance L along the axis for each revolution. The ray can be developed on to a plane to form a straight line, as shown in Fig. 13, and the length of the path is equal to $(4\pi^2 r^2 + L^2)^{\frac{1}{2}}$, so the condition for constancy of path length becomes:

$$n(r)(4\pi^2 r^2 + L^2)^{\frac{1}{2}} = n(o)L$$

$$\text{i.e. } n(r) = \frac{n(o)}{\left(1 + \left(\frac{2\pi r}{L}\right)^2\right)^{\frac{1}{2}}}$$

$$= n(o)(1 + \alpha^2 r^2)^{-\frac{1}{2}} \tag{12}$$

Equation (12) is not identical with Eq. (10), although both tend towards Eq. (9) in the paraxial region. The ideal medium for skew rays is not therefore ideal for meridional rays, and it has been shown that this will restrict the resolution achievable in this system when used as an

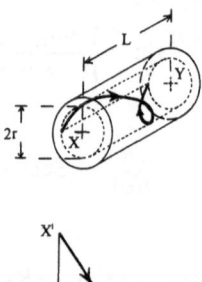

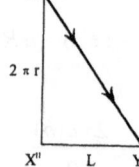

Fig. 13   The development of a helical skew ray (upper) on to a plane (lower) in a graded index fibre.

imaging system. The resolution of the system is strongly dependent on the ratio of fibre diameter to fibre length and increases as this ratio decreases, i.e. as the paraxial region is approached. If this ratio is 100:1, the system will transfer about 1000 bits of information (30 × 30 resolved spots). The effect of skew rays is not quite so serious in the case of an optical waveguide, although their influence must be taken into account in the design of the fibre.

This type of fibre is made using a technique involving diffusive ion exchange, and consists of immersing a glass rod in a bath of molten salt so that the heavy ions in the glass are replaced by the lighter ions of the salt, through a diffusion process. This exchange will take place more quickly in the outer layers of the rod and results in a refractive index gradient across the rod in which the refractive index in the outer zones is less than that at the centre of the rod. This variation will be a monotonically decreasing function of the radius.

REFERENCES

1. N. S. Kapany and R. J. Simms, *J. Opt. Soc. Am.* **55**, 963 (1965)
2. E. Snitzer, *Phys. Rev.* **7**, 444 (1961)
3. C. K. Koester and E. Snitzer, *Appl. Opt.* **3**, 1182 (1964)
4. W. A. Shierclift and R. C. Jones, *J. Opt. Soc. Am.* **39**, 912 (1949)
5. G. T. Reynolds and P. E. Condon, *Rev. Sci. Instr.* **28**, 1098 (1957)
6. S. E. Miller, *Bell System Tech. J.* **44**, 2013 (1965)
7. S. Kawakami and J. Nishizawa, *IEEE Trans. Microwave Theory Tech.* **16**, 814 (1968)
8. E. G. Rawson, D. R. Herriott and J. McKenna, *Appl. Opt.* **9**, 753 (1970)

CHAPTER 11

Fibre Optics in Medicine

Historically, the medical field was the first to exploit the advantages offered by the use of optical fibres. In fact, the use of internal reflection as a means of guiding light was employed in the late 1930's to provide illumination for simple medical inspection instruments. In these instruments, the light from a filament bulb was guided to the inspection area by internal reflection along a polished plastic probe. The dielectric interface was formed between the plastic and air, as illustrated in Fig. 1. The plastic used was methyl methacrylate and the probe could be shaped to ease entry, or to perform some other function, e.g. retraction; some typical shapes are illustrated in Fig. 2. This form of illumination is still used in a variety of medical instruments but, since the reflecting interface is the probe surface, the efficiency is adversely affected by scratches and liquid films on this surface. Because of this, the probe is kept relatively short and with a large cross-section, so that the number of reflections is small. This restricts the range of instruments to those which do not require a large penetration and enter a relatively large cavity, and are normally used for oral or anal inspection. The main impetus for the use of fibre optics in medical instruments came from considerations of patient safety and comfort.

I. ILLUMINATION OF MEDICAL INSTRUMENTS

In conventional instruments the required illumination is provided by means of a small filament bulb situated at the distal end of the instrument. The electrical power to the bulb is supplied through leads which enter the instrument at the eyepiece end, and are normally contained in a metal tube which also holds the bulb. The entire illumination system can therefore be easily removed from the instrument; such a system is illustrated diagrammatically in Fig. 3. There was an ever-

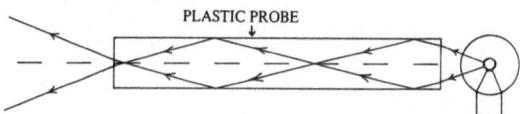

Fig. 1   Diagram illustrating the guiding of light from a lamp along a polished plastic probe.

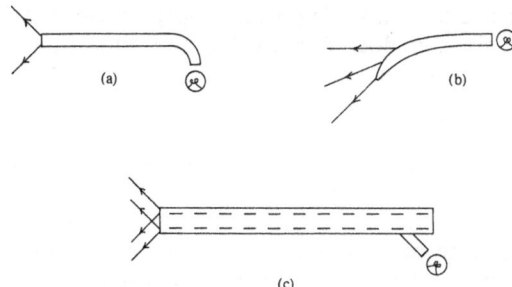

Fig. 2 Illustration of the shapes possible using clear plastic probes.:
(a) Rod bent at right-angle
(b) Retractor
(c) Hollow cylinder.

present danger of electrical shock owing to the live connections inside the patient plus the serious risk of bulb breakage. In addition, the wattage of the bulb had to be kept low to prevent burns, and consequently the light output from the bulb was small. A significant improvement over this system was achieved by using a light guide to pipe the light to the distal end of the instrument, from a lamp which was external to the patient, so that there were no live connections inside the patient. As an added advantage, heat could be filtered from the light at the source so that the possibility of burning the patient was eliminated.

The illumination system which has gained general acceptance is illustrated in Fig. 4. In this, the fibre-optics guide is in two parts: the first takes the light from the light source to the instrument and the second takes the light from a convenient access point on the instrument to the distal end. The two guides are coupled together mechanically so that the end faces of the fibre bundles are touching. The light source consists of a high-power filament bulb, means for condensing its output on to the input end of the light guide, a heat filter and some means of adjusting the output from the lamp, which can be either a

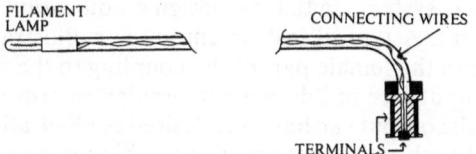

Fig. 3 Diagrammatic representation of a conventional endoscopic lamp, with a co-axial terminal arrangement

variable aperture or a variable power supply. Of these the former is to be preferred since it does not involve any alteration in the colour temperature of the lamp. In more sophisticated sources provision may be made for switching to a flash lamp to permit photography of the viewed area.

The light guide which couples the light source to the instrument is flexible and is typically 2 metres long, which allows the source to be positioned a convenient distance away from the patient. In operating theatres the source can also be placed some height above floor level, to minimize any explosion risk due to inflammable anaesthetics. This guide is sheathed in a flexible metallic trunking for protection, and the whole component is rendered water-tight either by using a sealed trunking or by covering the trunking with a plastic sleeve. The metal trunking is normally electrically insulated from the metal ferrules, to eliminate the presence of a conducting link to earth when diathermy is being carried out. Such guides are sterilized by chemical means, although certain designs are capable of being autoclaved many times with very little loss in transmission.

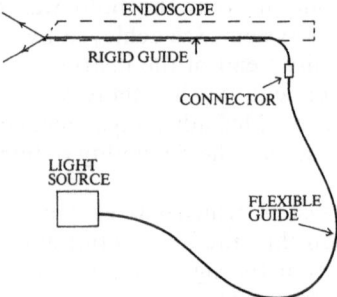

Fig. 4   Schematic lay-out of an endoscopic illumination system which uses fibre optics

The instrument guide is made to suit the particular instrument being used and normally takes the form of fibres packed into a fine metal tube bent to the required shape and polished at both ends. This directly replaces the light stem, which carried the bulb and connecting leads of the conventional system, so that no design modifications are required to the instrument. A typical instrument guide is illustrated in Fig. 5, which also shows the female part of the coupling to the flexible guide. Although the input face of this guide is circular, in order to match the flexible guide, the output can have any desired configuration; for example, those used with retractors are given a thin rectangular output to provide uniform light distribution with the minimum obstruction of vision. The output face of this light guide may also be angled to refract

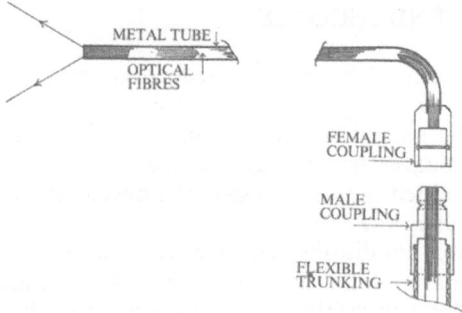

Fig. 5 Section through the rigid light guide used for endoscopic illumination. The male-female coupling to the flexible guide is also illustrated

the light beam away from the guide axis. The increase in available light can be seen from Fig. 6, which shows the variation in illumination with distance from the output face for both the fibre-optics and the conventional systems. This higher illumination has greatly increased the amount of information which can be obtained during the inspection, particularly with reference to colour changes in the tissues, and has widened the diagnostic uses of these instruments.

The use of fibre optics in instrument illumination has now spread to the dental field and a range of clip-on guides for dental tools is available using the same type of light source and coupling guides as in the medical case, although battery-powered versions are also available. These are valuable since they provide illumination at the point of interest and do not suffer from the disadvantage of the conventional dentist's lamp, namely that the light is obstructed by the dentist's head.

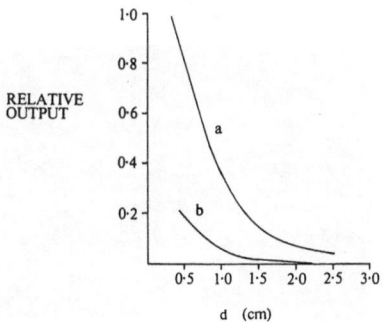

Fig. 6 Variation of the relative light outputs from (a) a fibre optics system, and (b) a conventional system, drawn as a function of distance (d) from the instrument.

II. MEDICAL ENDOSCOPES

The use of coherent bundles in medical endoscopy is perhaps the most widely known application of fibre optics, and illustrates the advantages to be gained by the correct utilization of this technology. In fact, it is probably true to say that the realization of the possible benefits to be gained in this area gave the development of fibre optics its initial impetus in the 1950s.

There are two conditions which have to be satisfied in an ideal internal examination and these are, first, that all areas within the cavity should be accessible, and, secondly, that the area which is viewed should be seen with adequate resolution and illumination. The classical rigid endoscope, comprising a train of lenses, fulfilled only the second of these conditions, and even then the illumination available was only marginally acceptable, for the reasons outlined in the previous Section. As was shown in that Section, the incorporation of optical fibres into the illumination system of such instruments has resulted in adequate fulfilment of the second condition. However, the accessibility to these instruments of the areas of interest is limited and skilled manoeuvring of the patient and instrument are required to achieve even this limited accessibility. Patient discomfort is high and certain vital areas are inaccessible, e.g. the duodenum in gastroscopy.

The incorporation of a flexible coherent bundle into the viewing section of such instruments gives an immediate increase in accessibility and a reduction in patient discomfort. The general optical arrangement is identical with that employed in the other forms of fibrescope, discussed in Chapter 8. The resolution of the overall system is adequate (Fig. 7) and comparable to that achieved by the classical system, the performance of which is degraded by the large number of lenses required, giving rise to ghost images and considerable aberrations.

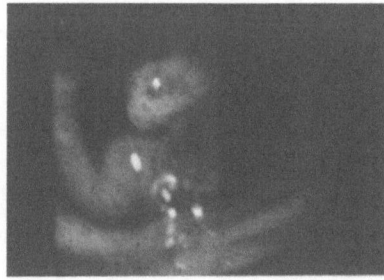

Fig. 7   Photograph of polypoid cancer in the stomach taken through a fibre optics gastroscope. (Courtesy of Olympus Optical Co.)

In general, the overall transmission of the fibre-optics system is higher than the equivalent lens train, where the Fresnel losses are high owing to the large number of lens elements (up to 50 of these being required in some instruments). The main criticism of the image transferred by the coherent bundle is that the image quality is degraded by the presence, in the image plane, of defects in the bundle, primarily broken fibres. These show up as black specks in the image and can be disconcerting to the viewer, although they do not obscure much information since their combined area is less than 1% of the total image area. The effect is therefore primarily a subjective one and is ignored by an experienced operator. However, it is because of this effect that the coherent bundles used in this application are of the mono-filament type, so that the effect of broken fibres is made subjectively less important. This was illustrated in Fig. 13 of Chapter 7 where a typical view has had 1% of its area blacked out in blocks corresponding to 20 μm single fibres and 60 μm multiple fibres. It will be noted that the subjective impression is that the larger blocks degrade the image to a greater extent. In addition, the bundles used normally have a specification which calls for virtually no breakages in the central zone of the bundle.

A. Types of Flexible Endoscope

The early development in the field of medical endoscopy took place in gastroscopy,[1] where there was a problem in that the duodenum was virtually impossible to examine with a rigid instrument. In addition, a fairly large bundle diameter could be tolerated in this field, which minimized the effect of broken fibres. A typical flexible gastroscope would be about 100 cm long in a flexible trunking about 1 cm in diameter, containing the coherent bundle and illuminating light guides. In addition, channels for biopsy, aspiration and insufflation are normally incorporated into this trunking. A modern instrument of this type is illustrated in Fig. 8.

In the early instruments there was no provision for remote steering of the distal end, and the positioning of the instrument was achieved by suitable manipulation of the instrument and patient, which was acceptable since patient discomfort was minimized. Later models have a remote steering capability similar to that discussed in Chapter 8 and shown in Fig. 9 which simplifies the task of locating the area of interest.

With the increased amount of light now available, still and cine photography is possible with these instruments, and in this connection, a Japanese instrument is of interest (Fig. 10). In this, a miniature camera is affixed to the distal end of the instrument and is adjusted to take a photograph of the area at which the coherent bundle system is looking. The camera shutter can be operated from the eyepiece end and the coherent bundle is used only to locate the area which it is desired to photograph. The resulting photograph obviously does not contain any defects due to the fibre optics and the system has the added advantage that the light falling on the film has not been attenuated by passage through another optical system.

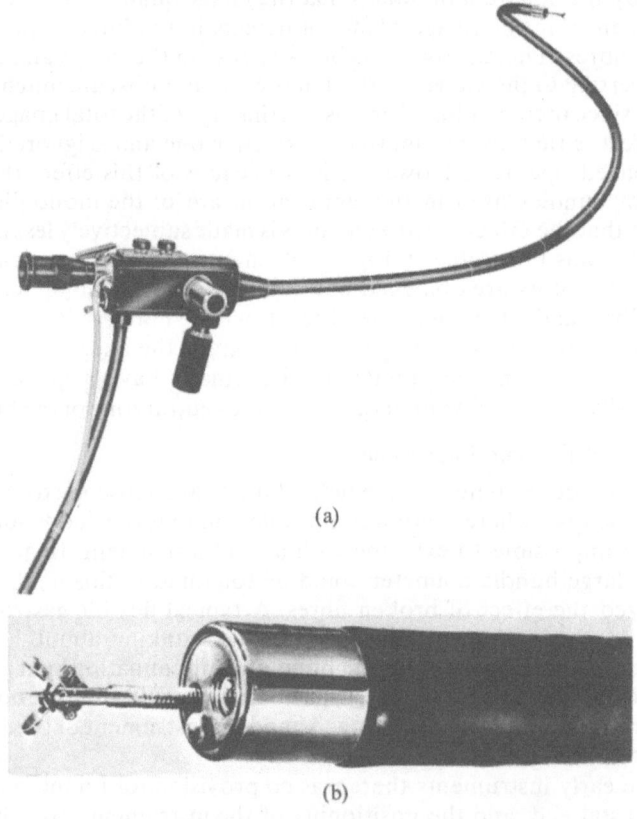

(a)

(b)

Fig. 8 (a) Photograph of a modern gastroscope which incorporates fibre optics for viewing and illumination
(b) Photograph of the distal end of the instrument depicted in (a), showing, in clock-wise order, biopsy forceps, illumination output (white disc), objective lens and air/water jet.
(Courtesy of A.C.M.I.)

As the development of coherent bundles progressed, the number of broken fibres in the bundles decreased so that bundles of smaller cross-section could be made with acceptable quality. This has led to the development of instruments with smaller cross-sections, such as cystoscopes and peritineoscopes. Future developments could include a small cardioscope which can be introduced into the heart through a

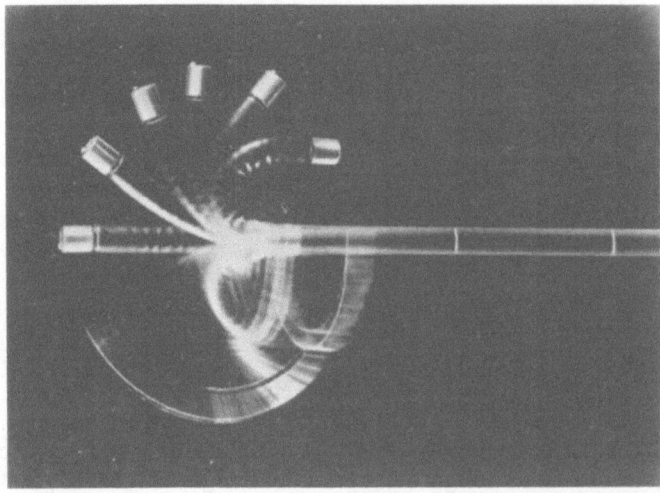

Fig. 9 Photograph showing the steering capability of a modern instrument (Courtesy of A.C.M.I.)

catheter tube, where the blood between the heart wall and the instrument can be displaced by filling a transparent balloon with saline solution to expand it against the wall. With the development of remote steering mechanisms, instruments have been developed which require to be steered around sharp bends and an example of this is a colonoscope, which has to negotiate the bends in the colon. The other design criteria are not stringent since the required length is 100 cm and outer diameters of 15 mm can be accommodated; a typical instrument is shown in Fig. 11.

As was indicated in Chapter 8, image conduits give resolutions and

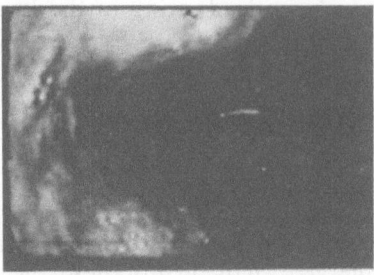

Fig. 10 Photograph taken inside stomach by the distal camera. The trunking of the instrument can also be seen. (Courtesy of Olympus Optical Co.)

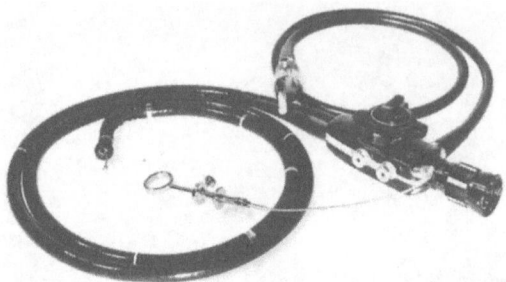

Fig. 11   Photograph of a modern colonoscope. (Courtesy of Olympus Optical Co.)

transmissions in excess of that achieved by a lens train so that even rigid endoscopes could be improved by the use of fibre optics. Image conduits have been incorporated in laryngoscopes, bronchoscopes and cystoscopes, although defects in the image tend to be greater at present in these than in flexible bundles. An interesting application of image conduit has been developed for the microscopic examination of living cells and tissue. It has been called a hypodermic probe[2] and consists of a piece of image conduit mounted inside a hypodermic needle with the distal end polished flush with the bevel of the needle. This is inserted into the tissue and the transferred image is that part of the tissue in contact with the end face of the image conduit, which can be examined under a microscope. Illumination means can also be incorporated into the probe or external illumination can be employed. The fibre diameters used are less than 5 μm and should ideally be around 1 μm for optimum performance. The image seen is distorted because the angled end-face of the image conduit has different magnifications in two mutually perpendicular directions, the higher magnification occurring along the greatest dimension of the end-face. This distortion can be eliminated using a suitable anamorphic lens system in the microscope. A schematic drawing of the hypodermic probe is shown in Fig. 12.

III. COHERENT BUNDLES IN MEDICAL INSTRUMENTATION

In addition to the direct application of the coherent bundle in endoscopy, as described above, two other applications of this type of component have emerged. In the first of these, the coherent bundle is used in medical training as a link between the instructor and the student. Mention has already been made of the skill required in handling

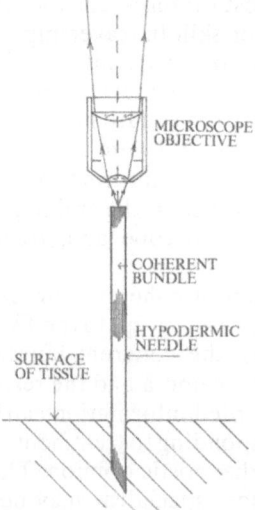

Fig. 12   Diagram illustrating the objective end of a hypodermic probe.

any type of endoscope, and this skill is difficult to transfer to a student via lectures or demonstrations. However, by incorporating a beam-splitter and flexible coherent bundle into the eyepiece of the endo-scope, as shown in Fig. 13, the student can observe with the instructor as the latter uses the instrument. In this way the student can quickly grasp the basic technique and becomes familiar with the salient features of the cavity which is being entered. In addition, the instructor can

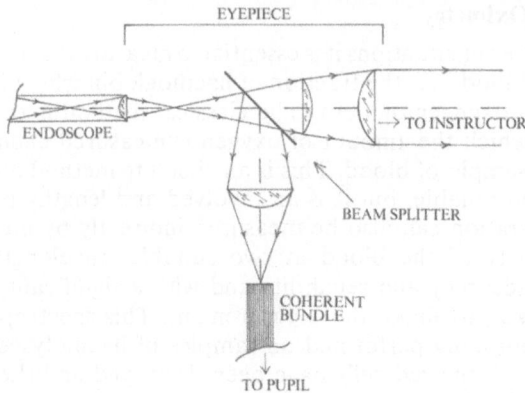

Fig. 13   Diagram of the beam-splitter used in the eye-piece of a training endoscope.

point out areas of diagnostic interest and those parts of the cavity which require particular care or skill in traversing.

An obvious extension to this training aid is to couple the coherent bundle to a television camera so that a whole class can be taught. In this case, the bundle is built from square multiple fibres in a 6 × 6 matrix, with an individual fibre size of 10 μm, and has a cross-section of 10 × 8 mm, which provides the same information content as the TV display. This has found extensive use in the training of students in the techniques of the operating microscope, and, to a lesser extent, in the training of dentists.[3]

The second application of coherent bundles in the field of medical instrumentation is similar in concept to the TV link, although the intention is different. In this, the coherent bundle is used to couple the endoscope to a television camera and the resulting output is recorded on video-tape. This recorded information can be used either as a check on how the patient is responding to treatment, or can be sent to another specialist for a second diagnostic opinion. This latter use would be of importance in areas where specialists may be separated by large distances, e.g. the United States of America.

IV. LIGHT GUIDES IN MEDICAL INSTRUMENTATION

Light guides can be used in medical instrumentation to perform functions more complex than a simple transfer of illumination. In this respect, the medical field is no different from any other area in which fibre optics is used, and light guides are found as part of opto-electronic systems in the area of medical measurement.

A. *In Vivo* Oximetry

In a number of situations it is essential to measure the oxygen saturation of the blood, i.e. the fraction of haemoglobin which is present in the form of oxyhaemoglobin. The classical method is the van Slyke method in which the amount of oxygen is measured chemically on a withdrawn sample of blood. This is an absolute method and therefore accurate and reliable, but it is an involved and lengthy process. The oxygen saturation can also be measured indirectly by measuring the optical density of the blood at two suitable wavelengths, with an equivalent accuracy and reliability and with a significant decrease in the time taken to make the measurement. This spectrophotometric method is normally performed on samples of haemolysed blood, i.e. blood in which the red cells have been destroyed and their pigments uniformly distributed in the serum, since this forms an isotropic medium in which the optical densities can be easily measured. However, any *in vivo* measurements must obviously be made with unhaemolysed blood which is a highly-scattering medium, since the red blood corpuscles have a different refractive index from the serum. In this case, the amount of light transmitted through the required path length is very small and the resulting measurements suffer from poor dynamic response and low accuracy.

In the case of haemolysed blood, Beer's law holds and the optical density, D, at any wavelength can be written as:

$$D(\lambda) = Wl \, (E_o \, (\lambda)C_o + E_r \, (\lambda)C_r) \tag{1}$$

where W is the total weight of haemoglobin/unit volume; l is the optical path length; E_o and E_r are the extinction coefficients of oxyhaemoglobin and reduced haemoglobin respectively; and C_o and C_r are the relative concentrations of oxyhaemoglobin and reduced haemoglobin respectively ($C_o + C_r = 1.0$). In the above formula the quantity C_o is required oxygen saturation.

It is found that E_0 and E_r are equal at a wavelength of 805 nm, called the isobestic wavelength. If this wavelength is λ_2 then, from Eq. (1):

$$Wl = \frac{D(\lambda_2)}{2E(\lambda_2)}, \text{ where } E(\lambda_2) = E_o(\lambda_2) = E_r(\lambda_2)$$

Therefore:

$$D(\lambda) = \frac{D(\lambda_2)}{2E(\lambda_2)} \{E_o(\lambda)C_o + E_r(\lambda)C_r\}$$

By measuring the density at a second wavelength, λ_1, the oxygen saturation can be expressed in the following manner:

$$C_o = a + b\frac{D(\lambda_1)}{D(\lambda_2)}$$

where a and b are constants which depend solely on the optical characteristics of the blood. In practice, λ_1 is chosen to be that wavelength at which the difference between E_o and E_r is a maximum, which occurs at 660 nm and gives the most accurate measurement.

For unhaemolysed blood, the following approximate relationships hold for the transmitted and back-scattered intensities at a particular wavelength:

$$I_t = I_o e^{-l(\epsilon + d)W}; \; I_b = I_o \frac{d}{2\epsilon}$$

where I_t and I_b are the transmitted and back-scattered intensities respectively; I_o is the initial intensity; ϵ and d are extinction and reflection coefficients respectively. The expression for transmitted intensity only holds if:

$$l(\epsilon + d) \gg 1, \text{ i.e. } I_t \ll 1$$

This means that the amount of light to be measured is extremely small and leads to the inaccuracy and slow response which was mentioned above.

The expression for back-scattered light, on the other hand, is valid when $\epsilon \ll d$, which is the case for blood, and reflected intensities of 10% of the initial intensity can be achieved at 805 nm, for $l > 3$ mm. This amount of light can be measured accurately and with the fast response time necessary to record the dynamic variations in saturation which is the attractive feature of *in vivo* oximetry.

Now, $\epsilon(\lambda) = E_o(\lambda)C_o + E_r(\lambda)C_r$, so the fraction of back-scattered light can be written as:

$$\frac{I_b(\lambda)}{I_o} = \frac{d(\lambda)}{2(E_o(\lambda)C_o + E_r(\lambda)C_r)}, \text{ from Eq. (1)}$$

For the two wavelengths selected previously, the following relation holds:

$$\frac{D(\lambda_1)}{D(\lambda_2)} = \frac{I_b(\lambda_2)d(\lambda_1)}{I_b(\lambda_1)d(\lambda_2)}$$

and it can be shown that the oxygen saturation, C_o, is given by an expression of the form:

$$C_o = a + b\frac{I_b(\lambda_2)d(\lambda_1)}{I_b(\lambda_1)d(\lambda_2)}$$

The variation of d with wavelength is illustrated in Fig. 14 for reduced haemoglobin and oxyhaemoblobin.

The required measurements can be made with a conventional optical system, by bringing the blood from the circulatory system to the instrument via a catheter and a cuvette. However, a much better approach is to transfer the measuring point to the circulatory system using a light-guide system,[4] which means that measurements can be made directly in any part of the circulatory system. The light-guide system is of the Y-guide type, discussed in Chapter 5, in which one branch is used to pipe light to the region of interest and the second branch collects the back-scattered light and transmits it to a photodetector. The system is encased in a catheter tube and can be positioned at the desired point by normal techniques. The entire oximeter is illustrated diagrammatically in Fig. 15, where it will be noted that the input to the Y-guide is chopped by a rotating disc which contains narrow band filters corresponding to the two wavelengths λ_1 and λ_2. The result-

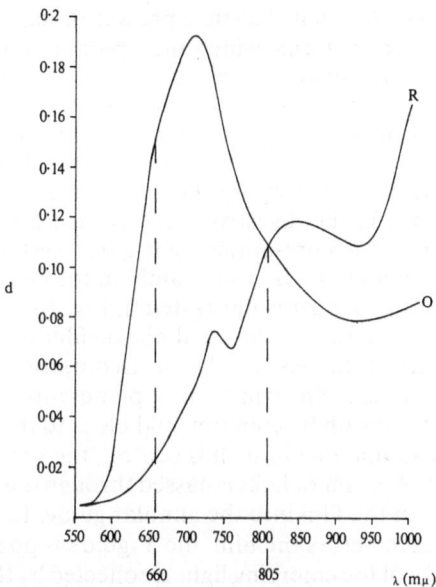

Fig. 14 The spectral variation of the diffuse reflection coefficient (d) for haemoglobin (R) and oxyhaemoglobin (O).

ing output is a measure of the relative amounts of back-scattered light at these wavelengths and can be processed electronically to give a measure of the oxygen saturation. Experimental evidence indicates that the relation given in Eq. (2) is valid to within a few per cent over the whole range of saturations.

B. Measurement of Pressure

A measurement of the fluid pressure at various points in the body can be a valuable diagnostic aid, particularly in the circulatory system

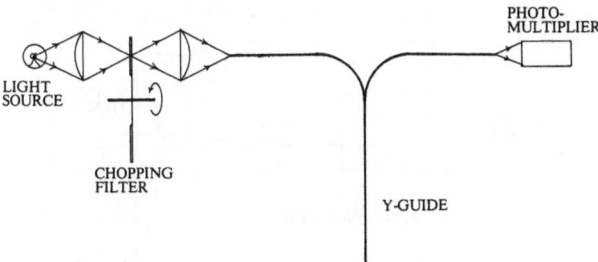

Fig. 15 Diagrammatic representation of an *in vivo* oximeter using fibre optics.

and the digestive tract. Small electrical pressure transducers are available and can be used in a unit which incorporates a transmitter: The whole unit is swallowed by the patient and its process through the tract is monitored by X-rays. Information regarding pressure variations within the tract is broadcast by the transmitter to be picked up and recorded by a receiver unit external to the patient. This system could not be used in the circulatory system, and gives rise to reliability problems when used in the digestive tract. A pressure transducer has been developed using a fibre-optics light-guide system, which gives increased reliability and can be used safely in the circulatory system.

The head of the transducer is illustrated in Fig. 16, where the transducing component is a thin aluminized plastic film which is stretched over the end of a small stainless-steel tube. In the absence of a pressure difference across it, the film will form a plane mirror. A Y-guide is positioned in the tube with its common end close to this film; the common end is made so that one branch is central, the second forming an annulus around it. A beam of light is passed through the central bundle and is reflected from the film into the annular guide. In the absence of a pressure difference across the film, the Y-guide is positioned so that approximately 50% of the emerging light is collected by the outer guide, and this is fed to a photo-cell which measures the amount of reflected light. The stainless-steel tube is fixed into a catheter so that there is no fluid connection between opposite sides of the film. If the pressure outside the transducer becomes greater than that inside, the film will be formed into a convex mirror and the light collected by the outer bundle will increase; conversely a reduction in pressure will decrease the amount of light collected.

The system can be calibrated to give a direct measurement of pressure although the resulting curve is non-rectilinear and tends to flatten out for large pressure differences. This can be overcome, and the accuracy of measurement increased, by converting the system to a 'null' method, which involves altering the internal pressure within the transducer until the measured light output indicates that there is again zero pres-

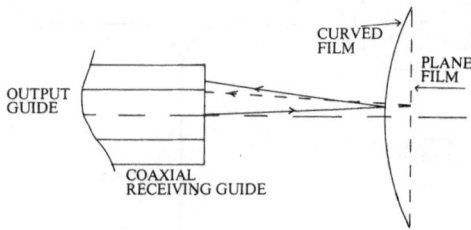

Fig. 16   Illustration of the principle of operation of the pressure transducer.

sure differential across the film; the applied pressure then equals the required pressure. This applied pressure is supplied to the catheter from outside the patient by an adjustable manometer. Further, by using this method the rectilinear portion of the response curve can be moved to any desired mean pressure enabling dynamic measurements to be taken about this mean value.

V. FIBRE OPTICS AND LASERS IN MEDICINE

The idea of transporting the power of a laser beam through an optical fibre bundle is a very attractive one, and could open up new avenues in bloodless surgery. However, there has only been a limited success in achieving this so far, the main problem being that the laser power which is useful medically also damages the fibre bundle. This damage affects primarily the output and input faces of the bundle and is therefore a coupling problem. The appearance of the damaged fibre is similar to that observed in samples of bulk glass. As yet, the underlying mechanism is not fully understood and, until this happens, there is little chance of being able to avoid this damage. It would appear that plastic fibres are less susceptible to damage than glass fibres, and light guides made with these are marginally acceptable in certain instances.

As in other applications, the use of fibre optics in conjunction with lasers would serve two purposes. First, a fibre bundle would be used to move the bulky, and possibly hazardous, laser system from close proximity to the doctor and patient to a more convenient position. A good example of this would be found in the laser opthalmoscope used for retinal welding. The laser system is at present housed in the handle of the instrument and yields a heavy instrument, which is also rather bulky. The incorporations of a light guide into the sytem would enable a much better instrument design to be achieved.

The second purpose which could be served would be to transfer the laser power via a fibre bundle to the point to be irradiated. A possible application in this case would be the removal of atheroma, which might be disintegrated by a powerful laser, from the circulatory system. In this connection, an experimental light-guide system was developed in which an attempt was made to overcome the laser damage at the input face of the light guide, which tends to be the more seriously damaged face. The light guide is shown diagrammatically in Fig. 17 and was made from tapered fibres; the input face contained the larger ends of the fibres. In this face, the fibre diameter was about 1.0 mm and tapered down to about 125 μm at the output end; twenty of such fibres were used which gave an input diameter of 5.0 mm and an output diameter of 0.6 mm. The input face was irradiated by light from an unfocussed ruby

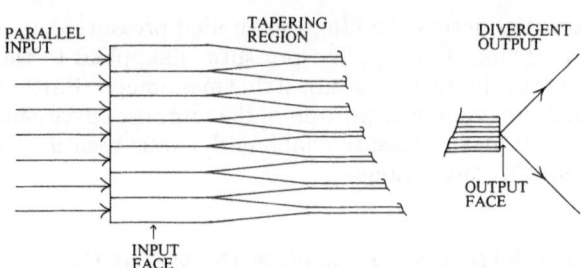

Fig. 17   Diagram showing the use of tapered fibres in conjunction with laser light.

laser, which did no damage to this face. The output beam was concentrated into a much smaller area by virtue of its passage through the tapered fibres, and a reasonable energy density was achieved at this output face. Damage did occur at the output face after a few shots and there was some evidence that particles of glass were ejected from the fibres at this end, which prevented further development.

Any components which are intended for use with laser light must not use an organic adhesive to hold the fibres, since this absorbs the laser radiation and reaches a very high temperature, which will cause damage to the fibres. Instead, the fibres at the ends of the components must be bonded by fusion of their sheaths, using heat and pressure, in a manner similar to that described in Chapter 7 for continuously-wound coherent bundles.

The use of lasing optical fibres in medicine has a somewhat limited potential, at present, since the main application would seem to be the replacement of the actual laser rod itself. Some work on this has been done, but it is difficult to see much advantage in this system, since the rod itself contributes little to the bulk of the system and is not hazardous in itself.

REFERENCES

1. B. I., Hirschowitz, L. E. Curtiss, C. W. Peters and H. M. Pollard, *Gastro enterology* **35**, 50 (1958).
2. D. F. Capellaro, N. S. Kapany and C. Long, *Nature* **191**, 927 (1961).
3. A. P. Hovnanian, Med. Electronics Conf. Proc., 1st, London (1960) p. 22.
4. N. S. Kapany and N. Silbertrust, *Nature* **204**, 138 (1964).

BIBLIOGRAPHY

The following references are given in addition to those quoted at the end of each chapter. These are grouped under the various chapter headings, plus a general heading which includes introductory and review articles.

GENERAL

1. Ballantine, J. and W. Allan. Fibre Optics, Science J. *1*, 73–9 (1965).
2. Coleman, K. R., Practical Aspects of Fibre Optics, Instn. Elect. Engrs. Conf. Rep. Ser., No. 5, 363 (1963).
3. Cope, J. D., and H. E. Brown, Fibre Optics Applications and Fabrication Techniques, S.P.I.E. Journal *6*, 6–8 (1967).
4. Courtney-Pratt, J. S., Some Unconventional Methods of High Speed Photography, In: Intern. Congr. High Speed Phot. Proc. 5th, Washington D.C., 1960. 1962, p. 197–226.
5. Drougard, R., and R. J. Potter, Fibre Optics, In: Advanced Optical Techniques, (A. C. S. van Heel, Ed.) Amsterdam, North Holland Publishing Co., 1967, pp. 399–433.
6. Fuller W. D., Optoelectronics – New Technology for Product Designers, In: IEEE – Product Eng. & Production Group Conf., Proc. 8th, 1964, p. 38–46.
7. Gallagher, T. J., Fibre Optics – A New Tool for Industry, IEEE Trans. Industr. Electronics, *IE 10*, 57–61 (1963).
8. Hansell, C. W., Picture Transmission, U.S. Patent 1,751,584 (1930).
9. Harvard University, Graduate School of Business Management, Fibre Optics, Management Reports 1964.
10. Hicks, J. W., and P. Kiritsy, Fiber Optics, Glass Industry, *43*, 193–6 (1967).
11. Hopkins, R. E., and R. J. Potter, Fiber Optics, In: Symposium on Cinefluorography, Proc. 1st, University of Rochester, 1958, p. 228–43.
12. Kapany, N. S., Electro-Optical Systems Using Fibre Optics, Optica Acta *7*, 201–17 (1960).
13. Kapany, N. S., Fiber Optics, Sci. American *203*, 72–81 (Nov. 1960).
14. Kapany, N. S., Fiber Optics and the Laser, Ann. N.Y. Acad. Sci. *122* 615–37, (1965).
15. Kapany, N. S., Role of Fiber Optics in Photography, Japan. J. Appl. Phys. *4*, Suppl. 1, 312–22 (1965).
16. Kapany, N. S., Role of Fibre Optics in Ultra-High-Speed Photography, J. Soc. Motion Picture and Television Engrs. *71*, 75–81 (1962).
17. Krolak, L. J., W. P. Siegmund, and R. G. Neuhauser, Fibre Optics A New Tool In Electronics, J. Soc. Motion Picture and Television Engrs. *69*, 705–10 (1960).
18. Krolak, L. J., Fiber Optics and Its Use in Electro-Optical Devices, In: *Light and Heat Sensing* (H. J. Merrill, Ed.) 6th AGARD Avionics Panel Meeting, Paris, 1962. London, Pergamon Press, 1963, p. 297–310.
19. Novotny, G. V., Fiber Optics for Electronics Engineers, In: *Optoelectronic Devices and Circuits* (S. Weber, Ed.) New York, McGraw-Hill Book Co., 1964, p. 9–14.
20. O'Brien, B., Physics in the Optical Industry, Physics Today *13*, 52–6 (Jan., 1960).
21. Progress Report on Fiber Optics Technology, Illum. Eng., *60*, 24–7 (1965).
22. Siegmund, W. P., Fiber Optics, In: *Applied Optics and Optical Engineering*, Vol. 4, (R. Kingslake, Ed.) New York, Academic Press, 1967, p. 1–29.
23. Siegmund. W. P., Fiber Optics, Glass Industry, *41*, 502–4 (1961).
24. Siegmund W. P., Fiber Optics – Principles, Properties and Design Considerations, In: *Light and Heat Sensing* (H. J. Merrill, Ed) 6th AGARD Avionics Panel Meeting, Paris, 1962. London, Pergamon Press, 1963, p. 265–96.

25. Siegmund W. P., Unconventional Fiber Optics, In: Image Intensifier Symposium, Proc., 2nd. For Belvoir, Va., 1961, p. 159–66.
26. Snitzer, E., Some Properties of Fiber Optics and Lasers, In: *Optical Processing of Information* (D. K. Pollack *et al.*, Eds.) Baltimore, Spartan Books, 1963, p. 61–73.
27. Tiedeken, R., New Optical Design Units and Their Possible Applications in Photo, Cine and Television Engineering, Bild un Ton. *20*, 2–7 (1967).
28. Veinberg, V. B., Fiber Optics in the USSR, Optiko-Mekh. Prom., No. 11, 48–51 (1967).
29. Veinberg, V. B., Use of Fiber Optics in Photography, Zh. Nauch. Priklad. Fot Kine. *5*, 370–6 (1960).
30. Weichel-Moore, E. J., and R. J. Potter, Fibre Optical Properties of Ulexite, Nature *200*, 1163–5 (1963).
31. Young, A. F. B., and J. G. C. Steel, High-Speed and Ultra- High- Speed Photography in Electrical Engineering, J. Phot. Sci. *5*, 112–20 (1957).

CHAPTER TWO – FIBRE OPTICS BASIC THEORY

32. Harrick, N. J., Electric Field Strengths at Totally Reflecting Interfaces, J. Opt. Soc. Am., *55*, 851–7 (1965).
33. Jones, A. L., Coupling of Optical Fibers and Scattering in Fibrers, J. Opt. Soc. Am. *55*, 261–71 (1965).
34. Kapany, N. S., Fiber Optics. V. Light Leakage Due to Frustrated Total Reflection, J. Opt. Soc. Am. *49*, 770–8 (1959).
35. Richter, W., The Aperture-Limitation Function for Curved Optical Fibres, Optik *23*, 517–22 (1965–66).
36. Richter, W., The Black Band Effect and Its Applications, Optik *23*, 436–42 (1965–66).
37. Richter, W., and H. Zschaeck, Radiation Characteristics of Optical Fibers and Cylindrical Rods, Optik, *25*, 182–96 (1967).
38. Richter, W., The Transfer of Energy Through Stretched and Bent Glass Fibres, Optik *23*, 589–95 (1965–66).
39. Sawatari, T., and K. Sayanagi, The Angular Distributions and Patterns of The Light Emitted from Glass Rods, Oyo Butsuri *34*, 214–6 (1965).
40. Smith, A., Glass Fiber Condensing Systems, M.S. Thesis, University of Rochester, 1957.

CHAPTER THREE – THE OPTICAL FIBRE

41. Hioki, R., and T. Suzuki, Coherent Light Transmitted Through Optical Fiber, Japan. J. of Appl. Physics, *4*, 817 (1965).
42. Spitz, E., and A. Werts, Transmission of Images Along an Optical Fibre, C. R. Acad. Sci. B *264*, 1015–8 (1967).
43. Werts, A. Propagation of Coherent Light in Optical Fibers, Onde Elect. *46*, 967–80 (1966).

CHAPTER FOUR – NON-COHERENT BUNDLES – MANUFACTURE AND PROPERTIES

44. Brown, R. G., Plastic Fiber Optics, I. Transmission of Ruby Laser Radiation, Appl. Optics, *6*, 1269–70 (1967).
45. Caulfield, H. J., and J. L. Harris, Light Pipe Holography Appl. Optics *6*, 1272 (1967).

46. Gilleo, M. A., Radiation Transfer by a Light Pipe between Media with High Indices of Refraction, Appl. Optics, *3*, 765–7 (1964).
47. Hager, T. C., and R. G. Brown, and B. N. Derick, 'Crofon' Plastic Fiber Optics, In: S.P.E. Ann. Tech. Conf., Vol. 13, 1967, p. 338–45.
48. Morita, E., and M. Takase, Angular Distribution of Light Emitted from the Optical Fuber Bundle, Japan. J. Apply. Phys. *6*, 414–5 (1967).
49. Reichel, W., Light Transmission by Means of Optical Conductors, Exper. Tech. Phys. *13*, 359–68 (1965).

CHAPTER FIVE – NON-COHERENT BUNDLES – APPLICATIONS

50. Field E., and E. F. Tubes, Use of Fiber Optic Light Pipes with Magnetically Driven Shock Tubes, Rev. Sci. Instrum. *31*, 64 (1960).
51. Goettelman, R. C., and J. K. Crosby, Optical Probe Techniques, Rev. Sci. Instrum. *35*, 1546–9 (1964).
52. Gorenstein, P., and D. Luckey, Light Pipe for a Large-Area Scintillator, Rev. Sci. Instrum, *34*, 196–7 (1963).
53. Hoffman, G. R., and D. C. Jeffreys, A High-Speed, Large-Capacity Fixed Store for a Digital Computer, In: *Optical Processing of Information* (D. K. Pollack *et al.*, Eds.) Baltimore, Spartan Books, 1963, p. 246–54.
54. Kapany, N. S., Electro-Optical Systems Using Fibre Optics, Optica Acta *7*, 201–17 (1960).
55. Kapany, N. S., J. N. Pike, Fiber Optics. IV. A Photo-refractometer, J. Opt. Soc. Am. *47*, 1109–17 (1957).
56. Kapany, N. S., Fiber Optics for Data Recording, In: Tech. Ass. Graphic Arts, Proc., 12th, Washington D.C., 1960, p. 95–101.
57. Kapany, N. S., and D. A. Pontarelli, Photorefractometer. 1. Extension of Sensitivity and Range, Appl. Opt. *2*, 425–30 (1963).
58. Kapany, N. S., and D. A. Pontarelli, Photorefractometer. II. Measurement of N and K, Appl. Opt. *2*, 1043–8 (1963).
59. Malyshev, G. M., and A. I. Ryskin, On the Possibility of Using Fiber Optics in an Arrangement Consisting of a Fabry-Perot Interferometer and an Electron-Optical Converter, Optics Spectrosc. *17*, 433 (1964).
60. Meade, R. A., Plastic Fiber Optics: New Design Tool of Auto Industry, SAE Journal *75*, 67–72 (Dec., 1967).
61. Neubert, P., Use of Lighting Pilot Cables for Object Lighting Bild und Ton *19*, 202–7 (1966).
62. Pontarelli, D. A., and J. E. Bridges, Fiber Optics – Passive Sensors for Nonelectronic Equipment, In: *Automation in Electronic Test Equipment*, Vol. 4 (D. M. Goodman, Ed.) New York, New York University Press, 1967, p. 27–133.
63. Potter, R. J., Component Evaluation for an Optical Data Processor, In: *Optical Processing of Information* (D. K. Pollack *et al.*, Eds.) Baltimore, Spartan Books, 1963, p. 168–86.
64. Robben, F., and Fraser, R., Fiber Optics for Spectroscopic Illumination, Appl. Opt. *10*, 1141, (1971).
65. Suzuki, T., Interferometric Uses of Optical Fiber, Japan, J. Appl. Phys. *5*, 1065–74 (1966).
66. Suzuki, T., Experimental Study of Interference in Optical Fibers, Japan. J. Appl. Phys. *6*, 348–55 (1967).
67. Veinberg, V. B., L. N. Ivanova, and D. K. Sattarov, Radiation – Energy Guides, Svetotekhnika *8*, 1–5 (July, 1962).

68. Verondini, E., A Light Pipe for Very Thin Large Area Scintillators, Nuclear Instrum. Methods *48*, 42–4 (1967).

CHAPTER SIX – COHERENT BUNDLES – BASIC THEORY

69. Donath, E., An Experimental Study of the Dynamics of Fiber Optic Bundles, Appl. Optics, *5*, 1319–24 (1966).
70. Drougard, R., Optical Transfer Properties of Fiber Bundles, J. Opt. Soc. Am. *54*, 907–14 (1964).
71. Kapany, N. S., J. A. Eyer, and R. E. Keim, Fiber Optics. II. Image Transfer on Static and Dynamic Scanning with Fiber Bundles, J. Opt. Soc. Am. *47*, 423–7 (1957).
72. Kapany, N. S., Optical Image Assessment, Nature *188*, 1083–6, (1960).

CHAPTER SEVEN – COHERENT BUNDLES – MANUFACTURE AND PROPERTIES

73. Borough, H. C., Some Considerations in Using Fiber Optics Image Couplers, S.P.I.E. Journal *2*, 171–4 (1964).
74. Chitayat, A., Enhancement of Images Transmitted through Fiber Optics, In: S.P.I.E. Image Enhancement Seminar, Proc. St. Louis, 1963, p. III–1–10.
75. Hopkins, H. H., and N. S. Kapany, Transparent Fibres for the Transmission of Optical Images, Optica Acta *1*, 164–70 (1955).
76. Kapany N. S., High-Resolution Fibre Optics Using Submicron Multiple Fibres, Nature *184*, 881–3 (1959).
77. Maus, G., Optical Fibrous Transmission of Images, VDI Z. *105*, 738–9 (1963).
78. Ohzu, H., T. Sawatari, and K. Sayanagi, Image Transfer Properties of Fiber Bundle, Japan. J. Appl. Phys. *4*, Suppl. 1, 323–9 (1965).
79. Sawatari, T., and K. Sayanagi, Image Transfer Properties of Optical Fiber Bundles, Oyo Butsuri, *34*, 207–13 (1965).
80. Wilcox, R. E., Diffraction by FIber Mosaics, Appl. Optics, 6, 582–3 (1967).

CHAPTER EIGHT – COHERENT BUNDLES – APPLICATIONS

81. Bender, H., Image Splitting Methods in High Frequency Photography, In: *Kurz-zeitphysik*, (K. Vollrath *et al.*, Eds.) Vienna, Springer-Verlag, 1967, p. 301–27.
82. Brouwer, W., and A. C. S. van Heel, Two-dimensional Coding of Optical Images, Optica Acta, *2*, 49–50 (1955).
83. Burroughs, E. G., and A. J. Kennedy, Electron Microscope Camera with Fiber Optic Output, Rev. Sci. Instrum. *37*, 771–2, (1966).
84. Capellaro, D. F., Fiber Optics in Data Display and Analysis, In: S.P.I.E. Photo-Optical Data Reduction Seminar, Proc. St. Louis, 1964, p. VII–1–10.
84a. Courteney-Pratt, J. S., A Fiber Optics Camera, In: Intern. Congr. High Speed Phot. Proc., 6th, The Hague, Netherlands, 1962. 1963, p. 30–40.
85. Courteney-Pratt, J. S., J. W. McLaughlin, E. C. Schramm, and H. Alberti, A Fiber Optics Camera for Recording Sequences of X-Ray Pictures, J. Soc. Motion Picture and Television Engrs., *71*, 585–90 (1962).
86. Day, R., and D. M. Krauss, Fiber Optics Yields a New Scanner Concept, Control Engng, *8*, 101–4 (Dec., 1961).

87. Doyle, R. J., A Scan-Conversion Tube Utilizing Fiber-Optics Photon Transfer, IEEE Trans. Electron Devices, *ED–10*, 410–6 (1963).
88. Gurevich, S. B., and V. A. Rabinovich, Optical Coding in the Reproduction of the Image in Television on Photography, Tekh; Kino i Televid. *10*, 38–44 (July 1966).
89. Holiday, C. T., Optical System for Line-Scan Television Satellite, In: S.P.I.E. Airborne Photo-Optical Instrumentation Seminar, Proc. Cocoa Beach, Fla., 1967, p. VII–1–8.
90. Hren, J. J., and R. W. Newman, Fiber Optics in Field Ion Microscopy, Rev. Sci. Instrum. *38*, 869–70 (1967).
91. Kapany, N. S., Fiber Optics. VIII. The Focon, J. Opt. Soc. Am. *51*, 32–4 (1961).
92. Kapany, N. S., Fiber Optics Coupling for Multistage Image Intensifiers, In: Image Intensifier Symposium, Proc., 2nd, Fort Belvoir, Va., 1961, p. 143–57.
93. Kilcoyne, M. K., Electronography and Image Intensification – A Comparison, In: Nat'l Aerospace Electronics Conf. Proc. 19th Dayton, Ohio, 1967, p. 259–60.
94. Korda, E. J., L. H. Pruden, and J. P. Williams, Scanning Electron Microscopy of P-16 Phosphor-Cathodoluminescent and Secondary Electron Emission Modes, Appl. Physics Letters *10*, 205–6 (1967).
95. Mueller, A. A., Considerations for Fiber Optic Application to Cathode Ray Tubes, S.P.I.E. Journal *6*, 44–8 (1967–68).
96. Potter R. J., and R. E. Hopkins, Fiber Optics and Its Application to Image Intensifier Systems, In: Image Intensifier Symposium, Proc., 1st, Fort Belvoir, Va., 1958, p. 91–109.
97. Potter R. J., and R. E. Hopkins, The Optical Coupling of a Scintillation Chamber to an Image-Intensifying Tube, IRE Trans. Nuclear Sci., *N.S.–7*, 150–8 (1960).
98. Stojanoff, C. G., A Transient Fiber Optics Probe for Space Resolved Diagnostics of Dense Plasmas, AIAA Journal *4*, 1766–72 (1966).
99. Uffen, R. W. J., Pictures through Fibers, Perspective *4*, 5–14 (1962).
100. van Heel, A. C. S., Optical Representation of Images Without Use of Lenses or Mirrors, Ingenieur *65*, 0.25–0.27 (1953).
101. Woodley, W. A., and D. Rogers, High Resolution and Fibre Optic Cathode Ray Tubes, Brit. Common. and Electronics *10*, 696–701 (1963).

CHAPTER NINE – WAVEGUIDE PROPERTIES OF OPTICAL FIBRES

102. Burke, J. J., Optical Switching with Fiber Optical Waveguides, In: Intern. Fed. for Inform. Process., Proc., New York, 1965, Vol. 2, p. 473–4.
103. Burke, J. J., Switching with Fiber-Optical Waveguides, J. Opt. Soc. Am. *57*, 1056–7 (1967).
104. Kane, J., Fiber Optics and Strain Interferometry, IEEE Trans Geosci. Electronics, *GE-4*, 1–10 (1966).
105. Kao, K. C., and G. A. Hockham, Dielectric-Fibre Surface Waveguides for Optical Frequencies, Proc. Instn. Elect. Engrs. *113*, 1151–8 (1966).
106. Kapany, N. S., Coherence and Radiation Effects in Dielectric Waveguides, In: Conf. on Coherence and Quantum Optics, 2nd, University of Rochester, 1966, p. 143–6.
107. Kapany, N. S., and J. J. Burke, Fiber Optics. IX. Waveguide Effects, J. Opt. Soc. Am. *51*, 1067–78 (1961).
108. Kapany, N. S., J. J. Burke, and K. Frame, Radiation Characteristics of Circular Dielectric Waveguides, Appl. Optics *4*, 1534–43 (1965).
109. Snitzer, E., Cylindrical Dielectric Waveguide Modes, J. Opt. Soc. Am. *56*, 1484 (1966).

110. Snitzer, E., Optical Dielectric Waveguides, In: Intern. Conf. on Quantum Electronics, Proc., 2nd, Berkely, 1961, p. 348–69.
111. Snitzer, E., Optical Dielectric Waveguides, In: Electromagnetic Theory and Antennas Symposium, Proc., Copenhagen, 1962, p. 903–6.
112. Snyder, A. W., Surface Waveguide Modes Along a Semi-Infinite Dielectric Fiber Excited by a Plane Wave, J. Opt. Soc. Am. *56*, 601–6 (1966).

CHAPTER TEN – MISCELLANEOUS TOPICS

113. Blincow, D. W., and J. R. Webster, Light Collection from Long Thin Scintillator Rods and Optical Coupling, IEEE Trans. Nuclear Sci., *NS-11*, 38–43 (June, 1964).
114. Goodman, D. M;, Nondestructive Testing of Electronic Circuits with Fiber Optic Scintillators, Vidicon Data Sampling and Pattern Recognition Displays, In: *Automation in Electronic Test Equipment*, Vol. 4, (D. M. Goodman, Ed.) New York, New York University Press, 1967, p. 287–301.
115. Hawkins, R. D., Vibrating Optic Fibers, A New Technique for Audio-Frequency Information Processing and Pattern Recognition, In: *Optical Processing of Information*, (D. K. Pollack, *et al.*, Eds.) Baltimore, Spartan Books, 1963, p. 187–98.
116. Kapany, N. S., and R. J. Simms, Recent Developments in Infra-red Fiber Optics, Infrared Phys. *5*, 69–80 (1965).
117. Koester, C. J., Laser Action by Enhanced Total Internal Reflection, IEEE J. of Quantum Electronics, *QE-2*, 580–4 (1966).
118. Koester, C. J. and C. H. Swope, Some Laser Effects Potentially Useful In Optical Logic Functions, In: Symposium on Optical and Electro-Optical Inform. Processing Technology, Proc. Boston, 1964. 1965, p. 253–67.
119. Koester, C. J., Some Properties of Fiber Optics and Lasers. Part B., In: *Optical Processing of Information* (D. K. Pollack, *et al.*, Eds.) Baltimore, Spartan Books, 1963, p. 74–84.
120. Paxton, K. B. and W. Streifer, Analytic Solution of the Ray Equations in Cylindrically Inhomogeneous Guiding Media. Part 1, Meridional Rays, Appl. Opt., *10*, 769 (1971).
121. Ibid. Part 2, Skew Rays, Appl. Opt., *10*, 1164, (1971).
122. Reiffel, L., and N. S. Kapany, Some Considerations on Luminescent Fiber Chambers and Intensifier Screens, Rev. Sci. Instrum. *31*, 1136–42 (1960).
123. Snitzer, E., Noedymium Glass Laser, In: Intern. Conf. on Quantum Electronics, Proc., 3rd, Paris, 1963, p. 999–1019.
124. Standel, R. R., and R. E. Hendrickson, Infrared Fiber Optics Techniques, Infrared Phys., *3*, 223–7 (1963).
125. Vlasov. M. K;, Yu. V. Terekhov, and V. T. Turov, Light Conducting Characteristics of Plastic Scintillating Fibers, Pribory i Tekhnika Eksperimenta, No. 6, 83–7 (Nov–Dec., 1966).
126. Waller, R., Spectron, J. Sci. Instrum. *41*, 261–2 (1964).

CHAPTER ELEVEN – FIBRE OPTICS IN MEDICINE

127. Behrend, J., Proctoscopic Photography, J. Soc. Motion Picture and Television Engrs., *75*, 655 (1966).
128. Benedict, E. B., Esophagoscopy, Gastrocopy and Peritoneoscopy, Gastroenterology, *42*, 171–4 (1962).
129. Brewington, H. H. and K. Stecher, An Improved Technique for Obtaining Cortical Photoelectric Plethysmograms, J. Appl. Physiol., *22*, 187–8 (1967).
130. Burnett, W., An Evaluation of Gastroduodenal Fiberscopes, J. Brit. Soc. Gastroenterology, *3*, 361–5 (1962).

131. Capellaro, D. F., N. S. Kapany and C. Long, A Hypodermic Probe Using Fibre Optics, Nature *191*, 927–8 (1961).
132. DeLaCroix, R. F., A Fiber Optic Pressure Transducer for Physiological Pressure Measurement, Ph.D. Thesis, Purdue University, 1966.
133. Enson, Y., W. A. Briscoe, M. L. Polanyi and A. Cournand, In Vivo Studies with an Intravascular and Intracardiac Reflection Oximeter, J. Appl. Physiol. *17*, 552–8 (1962).
134. Enson, Y., A. G. Jameson, and A. Cournand, Intracardiac Oximetry in Congential Heart Disease, Circulation *29*, 499–507 (1964).
135. Frommer, P. L., J. Ross, D. T. Mason, J. H. Gault, and E. Braunwald, Clinical Applications of an Improved Rapidly Responding Fiber Optic Catheter, Am. J. Cardiol. *15*, 672–9 (1965).
136. Fulton, W. F., The Fiberscope in Gastroscopy, Am. J. Gastroenterology, *38*, 290–8 (1962).
137. Gamble, W. J., P. G. Hugenholtz, M. L. Polanyi, R. G. Monroe, and A. S. Nadas, The Use of Fiber Optics in Clinical Cardiac Catheterization I. Intracardiac Oximetry, Circulation *31*, 328–43 (1965).
138. Goldman, J. A., S. Bereskin, and C. Shackney, Fiber Optics in Medicine, New Eng. J. Med. *273*, 1425–6 (1965).
139. Goldman, J. A., S. Bereskin, and C. Shackney, Fiber Optics in Medicine New Eng. J. Med. *273*, 1477–80 (1965).
140. Harrison, D. C., N. S. Kapany, H. A. Miller, N. Sibertrust, W. L. Henry, and R. P. Drake, Fiber Optics for Continuous In Vivo Monitoring of Oxygen Saturation, Am. Heart J. *71*, 766–74 (1966).
141. Hett. J. H., and L. E. Curtiss, Fiber Optics Duodenoscope and Ureterscope, J. Opt. Soc. Am. *51*, 581–2 (1961).
142. Hirschowitz, B. I., G. C. Luketic, J. A. Balint, and W. F. Fulton, Early Fiberscope Endoscopy for Upper Gastrointestinal Bleeding, Am. J. Digest. Diseases *8*, 816–25 (1963).
143. Hirschowitz, B. I., Endoscopic Examination of the Stomach and Duodenal Cap with the Fiberscope, Lancet *1*, 1074–8 (1961).
144. Hirschowitz, B. I., Endoscopic Photography Using Fiber Optics, J. Soc. Motion Picture and Television Engrs., *73*, 625–6 (1964).
145. Hirschowitz, B. I., A Fibre Optic Flexible Esophagoscope, Lancet *2*, 388 (1963).
146. Hirschowitz, B. I., Fibre Optics in Modern Medicine, Med. Biol. Illus. *15*, 224–9 (1965).
147. Hirschowitz, B. I., Gastroduedenal Endoscopy with the Fiberscope, Bull. Gastrointestinal Endoscopy *8*, 15–22 (1962).
148. Hirschowitz, B. I., Gastroduodenal Endoscopy with a Fiberscope – An Analysis of 500 Examinations, Surg. Clin. North Am. *42*, 1081–90 (1962).
149. Hirschowitz, B. I., Photography through the Fiber Gastroscope, Am. J. Digest, Diseases *8*, 389–95 (1963).
150. Hovanian, H. P., P. B. Brand, T. A. Brennan, and E. Watkins, Current Applications of Fiber Optics to Fluoroscopy, Med. Electronics Biol. Engng. *1*, 71–4 (1963).
151. Hovanian, H. P., J. S. Longo, P. B. Brand, T. A. Brennan, and A. J. Bower, Electro-optic Monitor and Fluoroscope, J. Am. Dental Assoc. *64*, 323–8 (1962).
152. Hovanian, H. P., Fibre Optic Dental Television Monitor and Fluoroscope, In: Med. Electronics Conf., Proc. 1st, London, 1960 p. 22–4.
153. Hovanian, H. P., H. F. McCarthy, and F. L. Rose, Fiber Optic Surgical Illuminator, Surgery, 52, 872–4 (1962).
154. Hovanian, H. P., H. F. McCarthy, and F. L. Rose, Fiber Optic Techniques in Proctovaginoscopy, Am. J. Surgery *108*, 101–4 (1964).
155. Hugenholtz, P. G., W. J. Gamble, R. R. Munroe, and M. L. Polanyi, The Use of Fiber Optics in Cardiac Catheterization. II. In Vivo Dye-Dilution Curves, Circulation *31*, 344–55 (1965).

156. Kapany, N. S., D. C. Harrison, N. Silbertrust, R. P. Drake, T. McLaughlin, and H. A. Miller, Fiber Optics Oximeter – Densitometer for Cardiovascular Studies, Appl. Optics *6*, 565–70 (1967).
157. Kapany, N. S., and N. Silbertrust, Fibre Optics Spectrophotometer for in Vivo Oximetry, Nature *204*, 138–42 (1964).
158. Lopresti, P. A., N. D. Scherl, L. Greene, and J. T. Farrar, Clinical Experience with a Glass-Fiber Gastroscope, Am. J. Digest. Diseases *7*, 95–101 (1962).
159. Lopresti, P. A., The Foroblique Fiberoptic Esophagoscope, Gastroint. Endosc. *31*, 20–1 (Aug. 1966).
160. Lopresti, P. A., A. Hilmi, and P. Cifarelli, The Foroblique Fiber Optic Esophagoscope, Am. J. Gastroint. *47*, 11–5 (1967).
161. MacDonald, H. Medical Applications of Fiber Optics, Biomed. Instrum. *1*, 15–7 (Oct., 1964).
162. Marshall, V. F., Fiber Optics in Urology, J. Urol. *91*, 110–4 (1964).
163. Marton, L., X-Ray Fiber Optics, Appl. Physics Letters *9*, 194–5 (1966).
164. Polanyi, M. L., and R. M. Hehir, In Vivo Oximeter with Fast Dynamic Response, Rev. Sci. Instrum. *33*, 1050–4 (1962).
165. Reynolds, W. E., S. Bazell, A. Brushenko, and D. A. Pontarelli, Fiber Optic Multiple Fiber Sigmoidoscope, S.P.I.E. Journal, *6*, 49–53 (1967–68).
166. Strub, I. H., Gastroscopy – A Re-evaluation, Am. J. Gastroenterology *40*, 75–81 (1963).
167. Turell, R., Fiber Optic Coloscope and Sigmoidoscope, Am. J. Surg. *105*, 133–6 (1963).
168. Wallace, F. J., Fiber Optic Endoscopy, J. Urol. *90*, 324–34 (1963).
169. Wallace, F. J., Fiber Optics, Hosp. Top. *43*, 109–11 (1965).
170. Wallace, F. J., New Diagnostic Aids through Fiber Optics, Am. J. Nursing *62*, 111 (1962).

Subject Index